Dedication

This book is dedicated to my wife, Robbi. Without her encouragement and support this book would probably not exist. It was her insistence that this instruction was badly needed. It was her willingness to carry an extra load and tolerate my frustration when I hit those inevitable writer's blocks and still kept encouraging me until the job was done.

Thank you for being there through it all. I love you.

Acknowledgements

Several people have been very instrumental in helping this book come into being. Their efforts deserve full recognition.

First, I'd like to thank my sister, Dr. Sue Atkinson for making the original suggestion to write this book and her support throughout the process.

Second, I would like to thank Dr. Bradly O'Mara who first showed me that such a book was actually a practical undertaking and that writing it would be a wonderful way to stay mentally sharp.

Third, I would like to thank John Barrow for his editing of the original manuscript. John, having been born, raised and educated in England, brought a greatly appreciated perspective on some of the explanations used for the concepts presented in this book.

Forth, I would like to thank Robert Countryman for all his help in getting this book published.

Fifth, to my sister-in-law, Shelly Thornton for her suggestions on the cover design.

Thank you, one and all.

Table of Contents

Introduction

If you are sixty years old or younger chances are very good that you grew up learning what was then called "New Math". Back in the 1960's the educational system decided to go away from teaching what we call conceptual math to teaching memorized math. The reason for this change has never been clear to me. Conceptual math was the method used to teach math to our young for over 6,000 years and resulted in all of the great advances in mathematics. Conceptual math is based on how the human mind works to solve problems. Learning conceptual math helps the mind develop its abilities to analyze and solve real world and abstract problems.

Memorized math is not based on how the mind works and does nothing to develop its abilities to think or solve problems. Memorized math is based on memorizing a method for solving different types of problems and requires very little mental exercise beyond that of memorization. But, you learned this method and it got you through school, right? Besides, how often have you had to use higher math skills since you graduated, right? So, why bother with this concept math, right?

Well, here are a few reasons you may find interesting:
1. In the real world you do use math whether you are aware of it or not, and with the kind of math you were taught you frequently get the wrong answer but either don't know it or can't figure out why or just give up.
2. A few years ago a group of concerned engineers contacted the federal department of education and told them that if this country didn't go back to teaching conceptual math that within the next ten years or so we would not be able to produce a homegrown scientist or engineer worthy of the title. That means that all of our brain people would have to be imported from other countries that still teach conceptual math. The federal government took their warning seriously and has begun strongly encouraging schools to go back to teaching proper math. This means that your children or grandchildren are probably now learning conceptual math. This is great, except when they come to you for help and

you're clueless about what to do.
3. Lifetime earning potential. There are many studies out there that show the difference between people who learned memorized math and those who learned conceptual math. The following chart is only one:

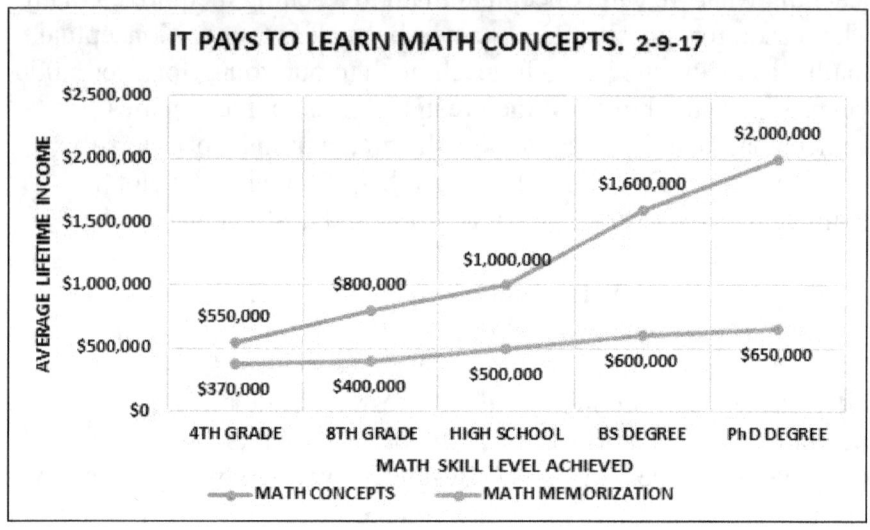

It's pretty obvious from the graph that learning the right kind of math pays dividends.

Because of the different kinds of math instruction, there now exists gaps between parents' understanding of math and what their children are learning. The purpose of this book is to help fill those gaps. It is written as much or even more for parents than for students. By using this book parents will be able to fill the gaps in their basic math education and then be able to help their children with their math education. The concepts are presented using real world examples and activities. I strongly encourage you to do the exercises with your children. It not only makes learning math more fun it also helps create some real quality time with you and your family.

Conceptual math is based on the real world we live in. It was this real world that first sparked men to create mathematics. For many years I was an electrical engineer working for a utility. As

6

part of my duties I was asked to teach our field engineers how to solve problems they would encounter with their jobs. I quickly discovered that very few of my students could understand what I was teaching them because it required a knowledge of trigonometry. I went to our educational department and asked if I could be allowed to develop and teach a one-day course in trigonometry. I was given permission and proceeded to develop the course.

The first time I taught the course it became apparent that I was also going to have to teach the fundamentals of algebra before my students could apply the trigonometry I was teaching them. The problem was that these people had not been taught conceptual math and had no idea how to apply math as a tool to solve problems. I quickly learned to adapt my classes to whatever part of the real world my students were familiar with. It one class we used the construction of a dog house to learn trig. In another class, we used quilting. The point is that we used the world around us to learn math and have fun at the same time.

Another reason for this book is the growing number of children who are being home schooled. For many people, home schooling is a wonderful alternative to standard education, however, there are some pitfalls. One particular pitfall is mathematics. Chances are very good that many parents were not taught conceptual math. Therefore, teaching conceptual math to their children can be a real challenge. This book is intended to help fill that gap.

I have spent some time pointing out the problems of memorized math, however, I would like to clarify something here at the beginning; anytime we learn something, some memorization is necessary. Without memorizing symbols, reading, writing and mathematics would be impossible. The good thing is that the world around us is full of ways to help us memorize what we need to in order to learn math. You will be reminded of this many times throughout the book.

Do not look at this book as a sole source of information but rather as a guide to understanding the way we learn and use mathematics. We are, after all, natural mathematicians. Everywhere you look we are surrounded by numbers and we use these numbers to deal with the world around us. We use numbers to tell time, to call our friends on the phone, to watch the TV, to drive our cars, to order food, to get paid, to tell the temperature just to mention a few. As you become more conscious of all the math that engulfs us you will, hopefully, become more aware that you don't need to be sitting in a classroom environment to learn math. One of the greatest places in the world to learn conceptual math is the grocery store. The next time you're there take a minute and help your children calculate which laundry detergent costs the least on a $/oz. basis. After a while, looking for opportunities to do math like this becomes a habit and it can be a lot of fun, not to mention useful.

In conclusion, as you read through the book I hope you will come to the realization that conceptual math is not some deep dark mysterious mind trick but rather the application of numbers to help the mind think the way it naturally does. We've been counting almost since the day we were born. It comes to us very naturally. Don't fight it! Go with it! You're going to be amazed at how much there is out there just waiting to be discovered by the simple use of math. Go...discover...have fun!

Chapter 1

The Basics
What, Why and How

What is mathematics? After you strip away the axioms, theorems, formulas, equations, methods, and so on and get right down to the basics, mathematics is counting. Does that sound too simple? All math is based around counting something. Over thousands of years we have found many new things to count and have devised incredibly complex and sophisticated methods for counting them but in the end, we are still just counting something.

What would you say if I told you most people are natural mathematicians? The standard response I get is a lot of shaking heads and looks that imply I'm not in touch with reality, but think about it for a minute. Math is counting and most of us start counting, in one form or another, from the time we're babies. As babies we start comparing how many fingers we have on one hand with the number of fingers on the other. Our minds can conceive of the number even though we don't have a name for it or a symbol to represent it. As we grow up we are continually finding more things to count, marbles, toys, pieces of candy and so forth. Then we got older and started school where most of us learned to dislike numbers and working with them because we were made to do a bunch of number crunching exercises that were pretty meaningless and boring and forced to memorize a bunch of stuff that was equally boring and meaningless.

So, why study math? If it was boring the first time won't it be boring this time? Let's start with the first question. I'm frequently told that there is no longer a need to study math because we have calculators and computers to do all of that for us. If that's true then why go to the gym and exercise? If I need to lift a heavy weight or move in a hurry don't we have machines for that? People exercise to strengthen their bodies so they'll be better able to handle the physical and mental stress of life. Mathematics is the gym for your mind. Studying math the right way strengthens your mind's ability to solve problems and deal rationally with the world around us. And don't forget, while a machine can crunch the numbers for you, it can't tell you how to setup the problem to get the solution. Remember I said that math is counting? Did you know that computers are only capable of doing two things, adding and comparing? All of the wonderful things they do are just the result of adding and comparing very quickly following some extremely clever programming.

Now, math doesn't have to be boring. Most people like playing games, so how do games work? Games consist of a playing environment, a game board, TV set, etc., playing pieces (real or virtual), a set of rules and a desired outcome. Now, within the game environment and using the playing pieces you are free to do whatever you can dream up to get to the desired result as long as you follow the rules. That's exactly how conceptual math works. The game environment is the real world, the pieces are numbers, the desired result is the answer to whatever problem we're trying to solve and the rules are the rules of how to manipulate our numbers. Mathematics is, in every sense, a game and it's a game that everyone can learn to play and enjoy at some level. Another wonderful thing about math is that, just like playing a game, there is always more than one right way to get the right answer. Don't ever think that you are going to be forced to learn one method to solve a given type of problem. What fun is that? When we do problems, we'll be looking at several different ways to solve each problem and encouraging you to find other methods on your own.

At this point I would like to emphasize that this book is about the beginning basics of math, but that doesn't mean it should be taken lightly or skimmed through. Its purpose is to build a solid foundation of comprehension before moving on to more advanced methods of problem solving. In my years of teaching math to technical people I have discovered that many of these people had weak basics in math. This resulted in their not understanding more advanced problem solving. Once we figured out their problem and spent some time relearning the basics they were able to move on to more advanced problems and actually start enjoying the whole learning process. Any good athlete, artist or musician will tell you that to get good at something you first have to master the basics. To stay good at something or even get better you have to keep practicing the basics.

Counting, Numbers and Symbols

Human beings count everything. We count so much that many of us tend to take it for granted but there is more to it than meets the eye. Let's say there is a circle on the floor and there are some marbles in it and each marble is a different color. If we ask someone to count how many marbles are in the circle they would probably tell us three, but don't be surprised if they hold up three fingers instead

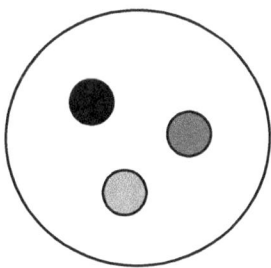

of saying three. They might also write out the word "THREE" or they may write the symbol "3". All of these answers are correct and they show the wonderful flexibility of the human mind and why there are sometimes challenges to learning.

Some people have difficulty differentiating one color from another. We say they are color blind. Some people have difficulty hearing tones as being different from each other and we say they are tone deaf. Some people have difficulty reading because their brains scramble the letters of words and we say they are dyslexic. Similarly, some people have difficulty associating the symbol "3" with the number of marbles in the circle. This has nothing to do with intelligence, in fact some very intelligent people throughout history have suffered from one or more of these afflictions yet have managed to lead very productive lives.

This brings us to learning styles. There are three basic learning styles, tactile, visual and verbal. Tactile people learn best when they can touch objects, feel them, move them around. Verbal people learn best when they are told something or when they can hear something explained to them. Visual people learn best when they can read words and see pictures or charts. There are several self-tests available on the internet that you can take to determine your most effective style of learning. I would strongly encourage taking at least one of these to find out your individual learning style. Knowing the best way for you to learn will be of tremendous value not only in learning math but in all areas of your life. It will also come in handy when you're trying to explain or teach something to someone else. If you can recognize their learning style then you can adapt your presentation to get your ideas across more effectively.

When it comes to counting, we use ten basic symbols, 0, 1, 2, 3, 4, 5, 6, 7, 8 and 9. It's extremely important that the mind learns to associate the correct number of objects with each symbol. This may seem trivial but over the years I have discovered that many people are weak in or sometimes even lacking this basic ability. If this is the case then the use of real marbles and flashcards can help train the mind.

Another interesting characteristic of the human mind is that it likes to group things together in smaller quantities then add these quantities together. Look at a pair of dice. The dots on each side of each die are arranged in a symmetrical pattern for balance and to make them easier to count. Let's say you role one die and get a 6. The dots are arranged in two rows of 3. When you first start learning to count you count each dot until you get to 6. The more you play with the dice the more your mind learns to make bigger groups and soon you see two rows of 3 dots. Eventually, you only see one group of 6 dots. Rolling dice and writing down the symbols for the number of dots on each die is a great way to learn to associate a number of objects with the proper symbol. Playing games that involve the use of dice and writing down the proper symbols for each die when you role is an effective and fun way to learn about counting. It's also a great way to start learning the basics of addition. In our previous example when you rolled a 6 and saw two rows of three dots your mind was adding 3 plus 3 to get 6.

So far what we've done works great as long as we don't have more than nine things to count. So, what happens when we do have more than nine things? To manage that we need to look at the bookkeeping of mathematics. There are many numbering or bookkeeping systems in use today and many more that have been used throughout history. The most common one used today is called the base 10 system. What does that mean? That means that we arrange larger numbers in columns by what is called powers of ten.

Let's say we take a small handful of toothpicks and toss them on the table. Now we count the toothpicks and find we have thirty-seven of them. Remember I said the human mind likes to group things and since we're using the base ten system let's group our toothpicks in groups of ten. After doing this we find we have three groups of ten toothpicks each with seven left over.

We arrange our numbers in columns. Ultimately, this makes working with them much easier. The first column on the right is for the leftovers, the seven toothpicks that couldn't be grouped together to make ten. We'll call this the one's column. The next column to the left is for how many groups of ten we have, we'll call this the ten's column, so we can write 37. This means that we have three groups of ten with seven left over.

What happens when we throw three more toothpicks on the table and put them in groups of ten? This time we end up with four groups of ten with none left over. Since we don't have any leftovers the symbol in the right-hand column will be 0. The symbol in the next column to the left will be 4 and our number is 40. By repeating this pattern, we end up with a bookkeeping system where no column can ever have a number in it great than 9.

Let's throw some more toothpicks on the table and group them in our usual groups of ten. This time we end up with nine groups of ten with nine left over. Again, following our bookkeeping system we write the leftovers down in the right-hand column, 9, then the number of groups of ten in the next column to the left, 9, and we have the number, 99. Now let's add one more toothpick to the pile. We've ended up with ten groups of ten or one hundred toothpicks. How do we write that number?

Since we have no leftovers the right-hand column is, 0. The next column is a problem since we have ten groups of ten but we can't have a number greater than 9 in any column. The solution is that we add another column to the left for groups of one hundred and call it the hundred's column. Now we can write our number by putting a 0 in the one's column, a 0 in the ten's column and a one in the hundred's column and get 100.

Now let's throw the rest of the box of toothpicks and the table. We know we have more than 100 toothpicks so let's start our grouping with groups of 100. This time we end up with 2 groups of 100 each with a bunch left over. Let's start making groups of ten. Now we end up with 3 groups of 10 with 6 left over. Now we can write our number by putting 6 in the one's column, 3 in the ten's column and 2 in the hundred's column and have 236 toothpicks.

This is how our math bookkeeping system works. The next time you see a number like 483 think to yourself that this is telling you that you have 4 groups of 100, 8 groups of ten and 3 left over. Now, just for practice, get five dice or more and start rolling them. Each time you roll them count all of the dots then write down the number. Keep doing this until this bookkeeping system is clear in your head and the numbers you write down actual mean something to you. Look around during the day and see what other things you can count and write those numbers down. After a while this system will seem very natural to you.

Exercises

1. Open up a snack bag of M&M's or Reese's Pieces and pour the candy onto a plate. Count how many pieces there were in the bag. Write the number down here _____.
2. The next time you go to the grocery store count how many checkout counters there are and write the number down on a piece of paper._____
3. When we make sidewalks we put groves in every few feet. The next time you're out for a walk count the number of groves for one block and write it down on a piece of paper._____
4. How many windows are in your house?_____
5. When writing numbers, if the column on the right is the one's column, the next column to the left is the ten's column and the next column to the left is the hundred's, what would we call the next column to the left? _____.
6. How many 10's are there in 100? _____.

Chapter 2

Putting the System to Work

Now that we have a system, let's take this thing for a test drive and see what it will do. Let's get out two dice and start tossing them. On the first throw we roll a 3 and a 5. In our bookkeeping system, we can write these numbers in the one's column like this:

1000's	100's	10's	1's	.1's	.01's	.001's
			3			
			+5			

We know by counting the dots that we have 8 dots. We can show this by writing:

1000's	100's	10's	1's	.1's	.01's	.001's
			3			
			+5			
			8			

This is called addition. Adding is just a faster way of counting. To get good at addition it's going to require that you do a little memorization at first. The more you play with the dice or count small groups of toothpicks the faster you'll start learning what the numbers add up to. For example, if you roll a 3 and a 6 you'll find they add up to 9. Once you start getting the hang of it, make yourself some flash cards and start practicing with them. The goal is to get to the point where you can add up small numbers, up to 4 plus 5, in your head quickly.

Now, let's take it up a level. Let's say you roll a 5 and 6. Write them in a column as before:

16

1000's	100's	10's	1's	.1's	.01's	.001's
			5			
			+6			

We know from counting the dots that this is eleven, which is one group of ten with one left over. In our bookkeeping system, we write the left over one in the 1's column and put a 1 in the 10's column to the left. Our answer looks like this:

1000's	100's	10's	1's	.1's	.01's	.001's
			5			
			+6			
		1	1			

Make sure to keep your columns going straight up and down. When it comes to math, neatness counts!

It's time to make up some more flash cards. This time create cards that show all of the combinations up to 9 plus 9. Once you have mastered these combinations you will have learned all of the addition tables up to 9 plus 9. Now during the day look around and find different things you can count and add to test yourself. For example, let's say you're standing on a corner and notice there is a fenced-in yard. Count the number of poles on one side and the poles on the next side then add them together. Get in the habit of carrying some paper and a pen or pencil with you so you can practice writing down the numbers in columns and also to make it easier to check your answers.

Let's look a little closer at our previous example of 5 plus 6. To really understand how our bookkeeping system works we need to make columns and show what we actually did in detail. Here's what the problem looks like with the columns drawn:

$$
\begin{array}{cc}
1 & \\
5 & \\
+6 & \\
\hline
1 & 1 \\
\end{array}
$$

What we did was an operation called **carrying**. When we added 5 and 6 we got 1 group of ten with 1 left over. We wrote down the left over 1 in the 1's column and then wrote down the number of groups of 10 at the top of the 10's column to the left. That's the grey 1 at the top of the 10's column. This is usually called carrying, in other words any groups of 10 we get when we add the 1's column we carry over to the top of the 10's column. We then added the number of groups of ten and wrote them at the bottom of the 10's column to get our answer. The 11 at the bottom is called the **sum**, so 5 added to 6 gives us a sum of 11. Let's do another problem to see why this way of thinking about adding is important.

Let's add 97 plus 36 showing our work in columns. Like before, the groups we are carrying will be written in grey:

$$
\begin{array}{ccc}
1 & 1 & \\
9 & 7 & \\
3 & 6 & \\
\hline
1 & 3 & 3 \\
\end{array}
$$

When we add the 1's column we get 13, which is 1 group of 10 with 3 left over. We write the 3 in the 1's column and carry the 1 group of 10 to the top of the 10's column. Next, we add the groups of 10, 1 plus 9 plus 3. Again, we get 13 which is 13 groups of ten or 1 group of 100 plus 3 groups of 10. We write the 3 groups of 10 in the 10's column and carry the 1 group of 100 to the top of the 100's column, then add the numbers in the 100's column and write that at the bottom to get the sum, 133. Remind yourself that this answer means we have 1 group of 100, 3 groups of 10 and 3, 1's.

By using our bookkeeping system this way we can keep track of our groups and it makes it easier to check our answers. This system is a very powerful tool for doing math. As we'll see later it is also very flexible. It will be used for subtraction and multiplication as well so take some time to get comfortable using it. Playing games with dice where you keep score is a great way to get good at using the system. Get together with some friends and play a game like Farkle or some similar game where you have a lot of opportunities to add numbers. You will probably want to have another piece of paper handy to add your numbers on. Remember to keep the columns straight. It can also be useful to have a calculator nearby to check your answers. If your answer doesn't agree with the calculator check your calculations again to see if you can find your mistake.

This brings up another question. Let's roll the dice again and let's say this time we get a 6 and a 3. When we write these numbers in a column does it make any difference which number we put on top? The answer is that it doesn't matter. If we count the number of dots on the dice it doesn't matter which one we count first, we're going to get a total of 9 either way. What if we had 5 dice, would the order make a difference then? By now it should be clear that the answer is no. When it comes to addition the order in which we add the numbers has no effect on our sum. There is a name for this property. It's called the **Commutative Law.** Later, we'll findout that the same law applies to multiplication.

It's fun to look for opportunities to use your addition skills. Suppose you have a couple of friends over and each of you has some cookies. You have some chocolate chip cookies, one friend has some ginger snap cookies and the other friend has some vanilla wafers. Have each person count their cookies then write down how many cookies each of you has and add them up to see how many cookies you have altogether. As you look around you'll soon discover there's no end to the chances to use your counting and math skills.

Another great opportunity for counting is when you're buying something. Money uses the same bookkeeping system but we're going to need to expand our system a little. Let's say the price of something is $4.17. That little dot in the number is called a

19

decimal point, so, where does it come from and how does it fit into our system?

In our system the smallest valued column we have used so far is the one's column, but what if we wanted to write a number that is less than one? That's where the decimal point comes in. Remember that each column going to the left is a larger group, one's, ten's, hundred's, etc. Now look at it starting with the hundred's column and going to the right. Each column represents a smaller group. After we get to the one's column, if we put in that decimal point and keep adding columns to the right, our pattern will continue. The first column to the right of the decimal point represents a group that is smaller than the one's. The next column to the right is a smaller group still.

If we look at that price again, $4.17. The 7 tells us how many pennies. The 1 in the next column to the left is telling us there is 1 group of 10 pennies. If we have 10 groups of 10 pennies each we have 1 hundred pennies or 1 dollar. The 4 to the left of the decimal point is telling us we have 4 dollars. As you can see our system hasn't changed. Dollars are in the one's column and numbers to the right of the decimal point are groups of pennies or parts of a dollar.

Let's say we want to buy a book for $5.27 and a candy bar for $1.44 (we won't include tax in these examples; that will come later) but we don't know if we have enough money to pay for it. All we have to do is add these two prices together to see. Let's use our columns:

$$
\begin{array}{c|c|cc}
 & & \overset{1}{} & \\
5 & . & 2 & 7 \\
1 & . & 4 & 4 \\
\hline
\end{array}
$$

6 . 7 1

Our total comes to $6.71, 6 dollars and 71 cents. As you can
20

see the decimal point really doesn't change anything we're doing, our system works exactly the same. The next time you go to the store have some paper handy and write down the prices of the things you are buying and add them up as you go. It also wouldn't hurt to take along a calculator to check your work. The store is a great place to exercise your math skills and you'll always know if you have enough money to pay for the things you're buying.

Remember to keep practicing your counting and adding. Challenge yourself to see how many things you can add in your head. This is great way to strengthen your mind and eventually you'll be amazed at how strong your math skills can be. Keep using the flash cards and practice with them whenever you've got few spare minutes. This is all about learning the basics and, like any activity, the better you are at the basics the better you'll be at the activity. If you have access to the internet do a search for math games and find some that are fun to play and spend time playing them. It will be time well spent.

As you gain experience you'll find there are other tips and tricks of the trade, so to speak, that can help you in different situations. For example, let's say you have to add a long column of numbers. One useful trick is to make small sums as you go. Let's add the numbers 5,8,3,9,2 and 7.

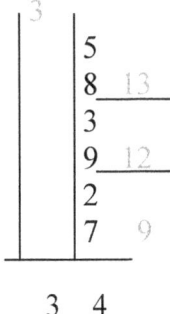

Here we've listed our numbers and at convenient points down the column we've drawn a line and added the numbers down to that line. These are called subtotals and are the numbers in grey. We then add the numbers in grey in the usual way and get 4 in the one's column with 3 groups of ten to carry to the ten's column. Adding the ten's column, we have 3 and our answer is 34.

Another way would be to add 5 and 8 to get your first subtotal, then add 3 to that subtotal then add 9 to that subtotal and so forth until you've added all the numbers.

```
   │   5
   │  +8
 1 │   3
   │  +3
 1 │   6
   │ + 9
 2 │   5
   │ + 2
 2 │   7
   │ + 7
 3 │   4
```

As you can see, there's more than one way to add numbers.

This is sometimes called long addition because the columns can get to be very long. Make a list of numbers yourself and give it a try. Don't be afraid to use a calculator to check your work, but remember, at this point in time a calculator should be used to check your work not do your work.

Next, we'll learn another very useful skill, subtraction, but first, how about some exercises?

Exercises

1. Write the following numbers in a column and add them up:
   5,9,4,2,8,3,1,6
2. Write the following numbers in a column and add them up:
   8,1,9,4,6,5,2,3. If you look closely you'll notice the numbers in this question are the same as question one only in a different order. This is one way of seeing the commutative law at work for yourself.
3. Add the following numbers:

1	2	3	4	5	6	7	8	9
+9	+8	+7	+6	+5	+4	+3	+2	+1

4. Suppose you're sitting at a restaurant and in your section you count 5 tables. Each table has 4 chairs around it. How many chairs are there?
5. On day you're walking down the street and since you're looking for different things you can count in order to practice your counting you start counting parked cars. After a few blocks your count reaches 12 parked cars. Each car has 4 wheels. What's the total number of wheels on the parked cars?

Chapter 3

Subtraction

Suppose you had 9 cookies and ate 3 of them. How many would you have left? You could count the cookies remaining or you could use subtraction. Subtraction, like addition, is another kind of counting only instead of combining groups we'll start with a group and then take a group away. We'll keep using the same bookkeeping system since it also works well for subtraction. And, just like addition, when you first start learning subtraction flash cards and dice can be very helpful. In this example, count three cookies and set them to the side. Now count how many cookies are left in the group and you'll see we have 6 cookies left.

Try using toothpicks, marbles or dots on a piece of paper and do a few more examples for yourself. In the case of dots on a piece of paper, make some dots on a piece of paper then use another piece of paper or your hand and cover a number of dots. Be sure to count how many dots you're covering up, then count how many dots are still showing. Eventually, you'll start to see a pattern. When you were learning addition, you came to learn that if you had 8 marbles and added 1 more you then had 9 marbles. With subtraction, you start with 9 marbles and when you take 1 away you have 8 marbles left. The more you practice your subtraction the more you'll see how the two are very much alike. You probably have a flashcard that shows that 5 plus 4 equals 9. In subtraction you'll have two flashcards, one that shows 9 minus 4 equals 5 and one that shows 9 minus 5 equals 4.

Just like addition, keep looking around for things to subtract. The more you do it the better you'll get at doing it. Remember when you started to learn a new computer game. The first few times you played it your score was probably pretty low. That was because you were just trying to learn the basics of how to play the game. The more you played the game the higher your scores became. That was because once you learned the basics you began applying them to winning the game. Learning math follows the same pattern, first learn the basics then keep applying the basics until you learn to use

24

them with confidence. The difference between the two is that the computer game is a make-believe world but math exists in the real world. You will always have a need to be able to add and subtract well.

Now, let's see how to use our bookkeeping system to do subtraction. Let's say you have 6 marbles and you want to give your friend 2. How many marbles will you have left? Just like in addition we'll use columns only this time we'll use a minus (-) sign in front of the group we're taking away to show we're subtracting:

$$
\begin{array}{r}
6 \\
-2 \\
\hline
4
\end{array}
$$

You can see that it looks very similar to addition. For now, always remember to put the group you're starting with on top, in this case 6, then write the group we're taking away, or subtracting, below. In order to learn the proper terminology, the top number, 6, is called the **minuend**. The number we're taking away, 2, is called the **subtrahend**. The answer is called the **difference**.

Now, let's take our bookkeeping system a little further. Let's say we have 24 toothpicks and we're taking away 3. Start by writing the problem the usual way:

$$
\begin{array}{r}
24 \\
-3 \\
\hline
\end{array}
$$

And, just like addition, we'll start with the 1's column. We take 3 away from 4 and we have 1 left. Then we go to the 10's column. Since we have nothing to subtract from our 2 groups of ten we can just bring the 2 down to our answer and we get:

$$
\begin{array}{r}
24 \\
-3 \\
\hline
21
\end{array}
$$

Now, let's do a problem where we need to subtract

something in the 10's column. We start with 36 cookies and eat 11 for desert. How many do we have left for desert tomorrow?

$$36$$
$$-11$$

In the 1's column we take 1 away from 6 and have 5 left. Now go to the 10's column and take 1 group of 10 away from 3 groups of 10 and have 2 groups of 10 left. That gives us our answer:

$$36$$
$$-11$$
$$\overline{25}$$

Now we know we have 25 cookies left for tomorrow. Now let's do a problem using the 100's column:

$$369$$
$$-155$$

Just like before we start with the 1's column and work our way to the left. 9 minus 5 is 4. 6 minus 5 is 1. 3 minus 1 is 2. Our difference is 214:

$$369$$
$$-155$$
$$\overline{214}$$

So far, our bookkeeping system is working very well, but let's look at another problem:

$$51$$
$$-28$$

When we go to subtract our 1's column we notice that we have to take 8 away from 1. How are we going to handle that? Think back to addition. When we added our 1's column and had more than 10 we carried that group of 10 to the 10's column. In subtraction, we're going to do the opposite, we're going to bring a group of 10 back into the 1's column. The process is called **borrowing**. Don't let the name mislead you. Normally when you borrow something you intend to give it back. In subtraction, we going to keep it. Here's how it works. In the 10's column of the minuend, the 5, we're going to take away 1 group of 10 leaving 4 groups of 10 and add that 10 to the 1's column of the minuend, the 1, giving us 11 in the 1's column. Now we can subtract 8 from 11 leaving 3 in the 1's column. In the 10's column we have 4 left in the minuend so we subtract 2 groups of 10 from 4 groups of 10 leaving us with 2 groups of 10 in the 10's column. That gives us a difference of 23 for our answer. Here's what it looks like using our bookkeeping system:

$$
\begin{array}{r}
4 \ \ 11 \\
\cancel{5}\,\cancel{1} \\
-2\ 8 \\
\hline
2\ 3
\end{array}
$$

As you can see our system still works. In fact, by using the system we can subtract VERY large numbers as easily as small numbers. Let's try one and see how it works.

$$
\begin{array}{r}
9\ 5\ 8\ 2\ 4\ 8\ 6 \\
-6\ 4\ 2\ 0\ 2\ 5\ 4 \\
\hline
\end{array}
$$

Starting with the 1's column and working left we have 6 minus 4 is 2, 8 minus 5 is 3, 4 minus 2 is 2, 2 minus 0 is 2, 8 minus 2 is 6, 5 minus 4 is 1 and 9 minus 6 is 3. Here's what it looks like:

$$
\begin{array}{r}
9,582,486 \\
-6,420,254 \\
\hline
3,162,232
\end{array}
$$

Notice the commas. They aren't necessary for our system to work but over time mathematicians have gotten into the habit of putting in commas every third column counting to the left from the 1's column to make large numbers easier to read. Remember the mind likes to think in groups and this is just another way of grouping large numbers but it has no effect on how our system works.

No matter how big the number is, or to put it another way, no matter how many columns we have, we always start with the column on the right and work each column to the left until we have our difference. Now let's subtract some big numbers where we have to borrow in the middle.

$$2\,8\,6\,0\,9$$
$$-1\,2\,7\,3\,7$$

Looking at this problem we can see that in the 10's column we're going to have to borrow, 0 minus 3, but notice that we're also going to have to borrow in the 100's column, 6 minus 7. So, starting in the 1's column we have 9 minus 7 which is 2. Now in the 10's column we have 0 minus 3. To do this we have to borrow a group of ten 10's from the hundreds column which leaves 5 groups of 100 in the 100's column and gives us ten groups of 10 in the 10's column. Now we can subtract 3 groups of 10 from ten groups of 10 giving us 7 groups of ten in the 10's column. That means that in the 100's column we are left with 5 minus 7, so, let's borrow 10 groups of 100 from the 1000's column. That gives us 15 groups of 100 minus 7 groups of 100 in the 100's column which is 8 groups of 100 and leaves us with 7 groups of 1000 in the 1000's column. 7 minus 2 leaves us 5 groups of 1000. In the 10,000's column we have 2 minus 1 which is 1. Here's what the problem looks like in bookkeeping form:

$$
\begin{array}{r}
15 \\
7\,\backslash\,10 \\
2\,8\,,6\,0\,9 \\
-1\,2\,,7\,3\,7 \\
\hline
1\,5\,,8\,7\,2
\end{array}
$$

The system still works. Now let's try a problem that's a little more practical. Let's say you go to the local department store to buy a candy bar and a pad of paper. The candy bar costs $2.35 and the pad of paper costs $5.25. You have $10. Do you have enough money to buy your friend a candy bar?

Remember our last problem about money when we were learning addition. Let's start this problem the same way. First, let's add together the cost of the candy bar and the pad of paper:

$$\begin{array}{r} \$2.35 \\ +\$5.25 \\ \hline \$7.60 \end{array}$$

Now we know how much our purchases will cost. Next, we subtract that from how much we have to spend:

$$\begin{array}{r} 9 \\ 0 \; 10 \; 10 \\ \$\cancel{1}\cancel{0}.0\,0 \\ -\$\;7.6\,0 \\ \hline \$\;2.4\,0 \end{array}$$

In the pennies' column, we have 0 minus 0 which is 0, but look at the next column to the left. Here we have 0 minus 6 which means we have to borrow from the next column to the left. Unfortunately, the next column to the left is also 0, so what do we do? Simple, go to the next column to the left which is a one. Here we borrow that 1, which leaves 0 in that column and becomes a 10 in the column to the right. We then borrow 1 group of 10 from this column, leaving 9 in this column and giving us 10 in the column to the right of that. Now we have 10 minus 6, which is 4, 9 minus 7, which is 2, and nothing left in the furthest column to the left. So, we will have $2.40 left. Since a candy bar costs $2.35 it looks like we can get our friend a candy bar, too. It's always good to share.

Notice that the decimal point is useful in knowing when we're dealing with quantities less than 1, but when it comes to using our system it really doesn't matter. We always start our addition or subtraction problem with the column furthest to the right and move left as we add or subtract.

Another useful thing we can do with our system is check the answers to subtraction problems by using addition. We can add the answer, $2.40, to the subtrahend, $7.60, and see if we get the minuend, $10.00:

$$
\begin{array}{r}
1 \\
\$\ 2.40 \\
-\$\ 7.60 \\
\hline
\$\ 10.00
\end{array}
$$

As you can see we just showed that 2.40 plus 7.60 equals the $10.00 we started with so our subtraction was right. There are other examples of subtraction and addition problems in the appendix for you to practice on. The answers are also there but try to work the problems first then check your answers.

Chapter 4

Negative Numbers

Let's look at another problem dealing with money. Once again, we are at the store and we've picked up a carton of milk, which costs $3.93, a loaf of bread which costs $2.49, a bag of marbles for practicing our math which costs $2.98 and a candy bar which costs $1.49. As before we have a $10.00 bill in our pocket. So, let's work this problem the same way we did before. First, add up what we want to buy:

$$
\begin{array}{r}
2\ \ 2 \\
\$3\,.\,9\,3 \\
\$2\,.\,4\,9 \\
\$2\,.\,9\,8 \\
\underline{\$1\,.\,4\,9} \\
\$1\,0\,.\,8\,9
\end{array}
$$

It's pretty obvious that our $10.00 won't be enough to buy everything, in fact simply by looking we can see we need another 89 cents to finish paying for everything. If we try to finish the problem like we did before by subtracting $10.89 from $10.00 our system seems to fall apart:

$$
\begin{array}{r}
\$10.00 \\
\underline{-\$10.89} \\
\end{array}
$$

If we try to borrow to do our subtraction we run out of rows, not to mention we'd get the wrong answer. So, what do we do?

We're going to start by introducing two new terms, the **POSITIVE** number and the **NEGATIVE** number. Back when we were learning addition we started out with a group of something, dots, marbles, toothpicks, etc. and then we had another group of something. When we added these two groups together we got one larger group. The symbol we used to show that we were adding was the + or plus sign.

When we learned subtraction we also started out with a group of something, but then we took away a group and ended up with a smaller group. The group we started with and the group we took away both represented positive numbers.

If you go back to the very first subtraction problem we did we had 6 marbles minus 2 marbles and we wrote it like this:

$$\begin{array}{r} 6 \\ -2 \\ \hline \end{array}$$

The 2 was the number of marbles we were giving our friend. Since these 2 marbles were going away, our original group of 6 was getting smaller, or, to say it another way, it was getting reduced by 2.

In mathematics signs become very important, but they can also be kind of tricky. When we were learning how to add we used a plus (+) sign. As just mentioned, when we were learning how to subtract we used a minus (-) sign. These signs are called **OPERATORS**. Addition and subtraction are called operations and these signs tell us what operation is to be done to work the problem. The tricky thing is that mathematicians used the same + and – signs to show if a number is positive or negative.

Let's take another look at that last money problem. We can't subtract $10.89 from $10.00, but we can turn the problem upside down and subtract $10.00 from $10.89. As we saw our difference is $.89. In mathematics, we attach a minus sign to it and say it is a negative number, -$.89. This means that we are short 89 cents of having enough money to buy what we went to buy. Another way of saying what we just did is combining a positive number with a negative number. The $10.00 is the positive number, that's the money we have, and the $10.89 is the negative number, that's the amount we want to take away. If we were to write this problem out using the number's signs it would look like this:

$$\begin{array}{r} -\$10.89 \\ \$10.00 \\ \hline -\$\ .89 \end{array}$$

32

Notice that there is no + sign in front of the $10.00. In mathematics, we like to keep the notations as simple as possible. **If there is no sign in front of the number we assume it is positive**.

Is there some way to use negative numbers in something other than money? Sure! Let's say we owe a friend 103 marbles. We dump the marbles out of our bag and count them to find we have 76 marbles. Since we have 76 marbles we'll make them the positive number. The 103 marbles we owe our friend we'll make the negative number since they'll be going away. Now let's combine our two numbers by putting the larger on top and subtracting:

$$\begin{array}{r} -103 \\ \underline{76} \\ -27 \end{array}$$

What this answer means is that we need to come up with 27 more marbles for our friend.

Negative numbers serve a very important role in all branches of mathematics, science, engineering, accounting and so on. Because they are so important we need to be sure we understand how to use them. We've seen that **when we combine a positive number with a negative number we put the larger number on top and subtract**. Our answer will have the same sign as the top number. What do we do if we have to combine two negative numbers? For example, from our previous problem we owe our friend 27 marbles. Now we find out we owe another friend 16 marbles. Since both of these groups of marbles are going away they both have negative signs, -27 and -16 and, since the size of the group of marbles going away is getting larger, we should add these numbers together. Here's what it looks like:

$$\begin{array}{r} -27 \\ \underline{-16} \\ -43 \end{array}$$

The -43 means we owe a total of 43 marbles and...we need to practice playing marbles!

As was mentioned before, working with negative numbers can be tricky so let's invent a machine to help us visualize the process more clearly. We're going to start with a machine that looks very much like a balance. First, we'll attach a rod to the table so that it stands straight up. Next, we'll attach a crossarm to the rod so that the center of the crossarm rests on the rod. At each end of the crossarm we'll attach a wheel with a groove in it like you would find in a pulley. Starting at the center of the crossarm we'll paint the part going to the right black and the part going to the left grey. When the paint dries we'll paint on some numbers. We'll put the number "0" at the center of the crossarm where the black and grey portions come together. Going right, at even spaces, we'll paint the numbers 1 through 9. Now we go back to the center of the crossarm and do the same thing going to the left only this time the numbers will have a minus (-) sign in front of them. All of the numbers will be white.

Now, let's put in a string that runs along the crossarm and through the wheels at each end. We'll make the string long enough that it hangs over at both ends. Next, we'll attach a small bucket at each end of the string. The bucket at the right end of the crossarm we'll paint black to match that end of the crossarm and paint the one at the left end grey so it match's that end of the crossarm. Now our machine is starting to look like an old-fashioned balance. Next, we'll attach a pointer to the string so that it points to "0" when the buckets are empty. Lastly, we'll attach special springs to the wheels at the ends of the crossarm so that our string won't just pull all the way through when we put some weight in one of the buckets.

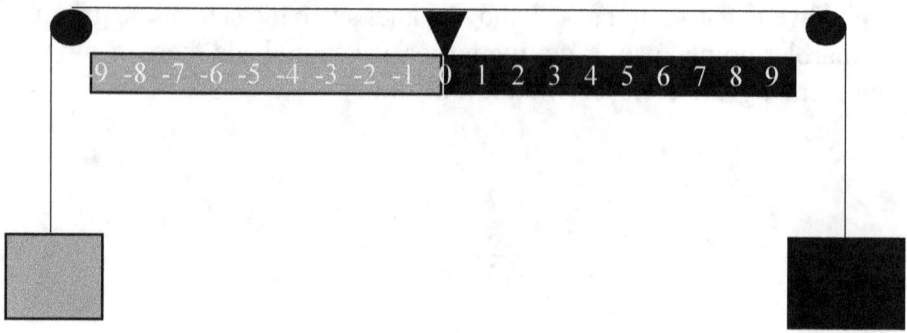

The last thing we need to do to make our machine useful is to create some special cubes. We'll make 20 cubes just to make sure we have enough. These cubes are all EXACTLY the same size and weight but half of them are black and half are grey. Now our machine is finished and ready to use. Let's figure out how to use it.

The right side of the machine, the black portion of the crossarm and black bucket, we're going to say are positive numbers. The left side of the machine, the grey portion of the crossarm and grey bucket, we're going to say are negative numbers. Putting cubes into either bucket we'll call addition, taking cubes out of either buck we'll call subtraction. Black cubes represent positive numbers and can only be put in the black bucket. Grey cubes represent negative numbers and can only be put in the grey bucket. Lastly, the weight of the cubes and the strength of the springs are such that when we put a cube in one of the buckets the pointer will move to the next number exactly. Let's try our invention and see how it works.

Let's add the positive numbers 2 + 3. First, we'll put two black cubes in the black bucket. The pointer moves right to the positive number 2.

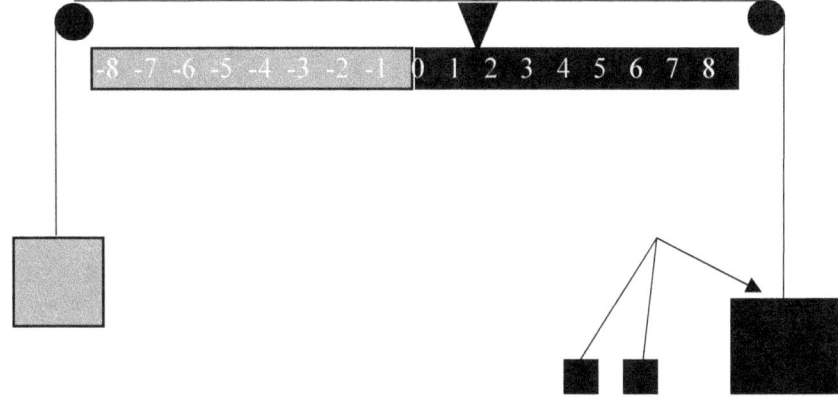

Now we'll put three more black cubes in the black bucket. The pointer moves three more spaces to the right stopping at the number +5.

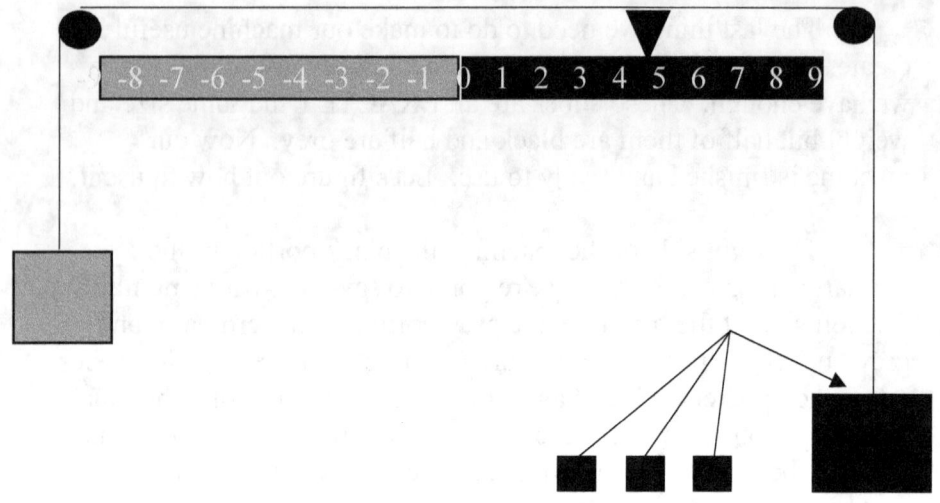

Our machine is telling us that 2 + 3 = 5 which we know is correct.

Now let's try a simple subtraction problem. We'll subtract 5 − 1. Both of these numbers are positive so to subtract 1 all we have to do is take out one black cube. When we do that the pointer moves left to the number +4. Our machine is telling us that 5 − 1 = 4 which we know is correct.

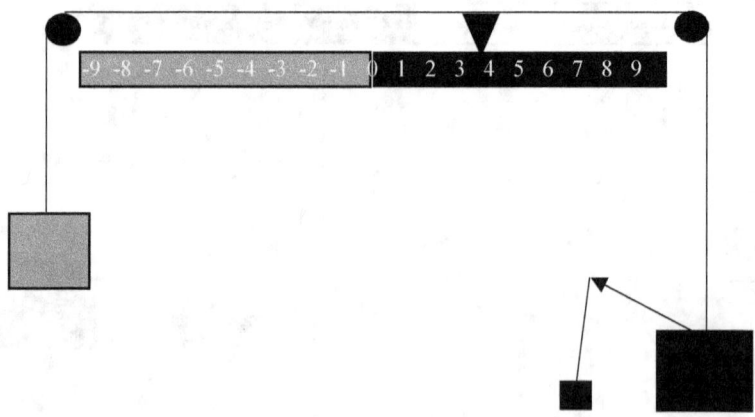

We'll put the black cube back in the bucket and the pointer goes back to pointing at the number +5.

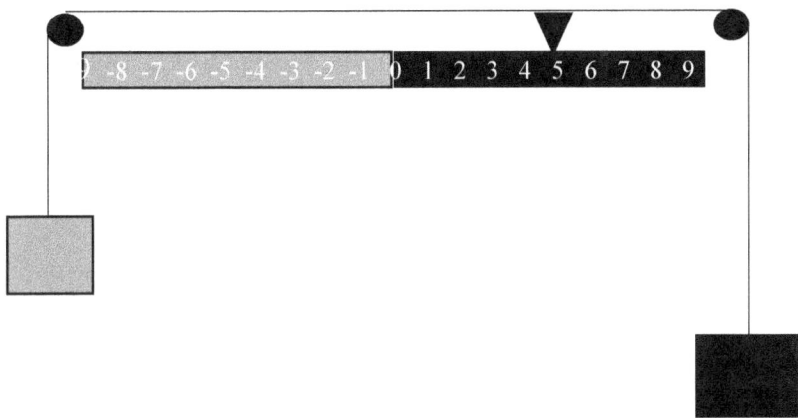

So, let's play with the negative numbers. We have five cubes in the black bucket and the pointer is pointing to the +5 on the positive side of the crossarm. Let's add one grey cube to the grey bucket. That's the same as adding a -1 or 5 + (-1). The pointer moves left to the number +4 again. Our machine is telling us that adding a -1 is the same as subtracting a positive 1. Can that be right? Yes, it is!

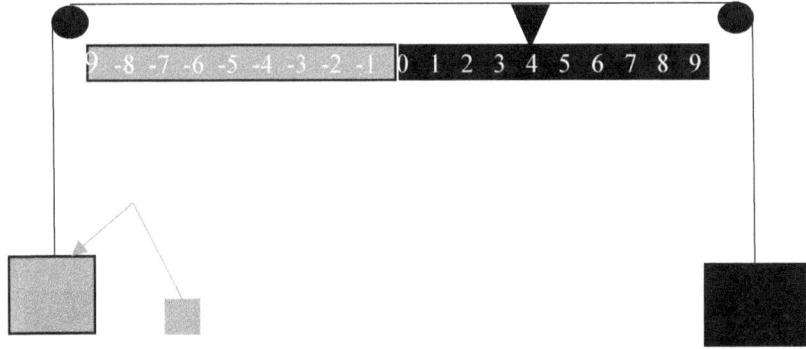

Let's do some more examples so we can get a clearer picture of how negative numbers work. We'll begin by taking all of the cubes out of the buckets so the pointer is at 0. Now we'll put 4 grey cubes in the grey bucket, that's the same as starting with the number -4.

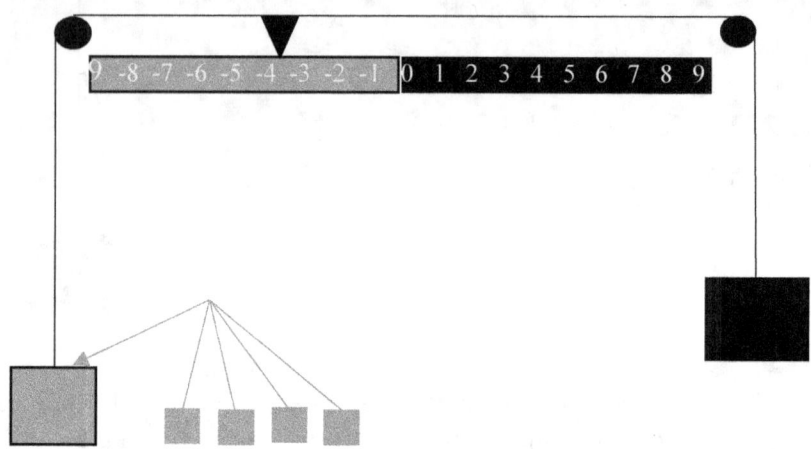

Next, we'll put 2 black cubes in the black bucket, that's the same as adding a positive 2.

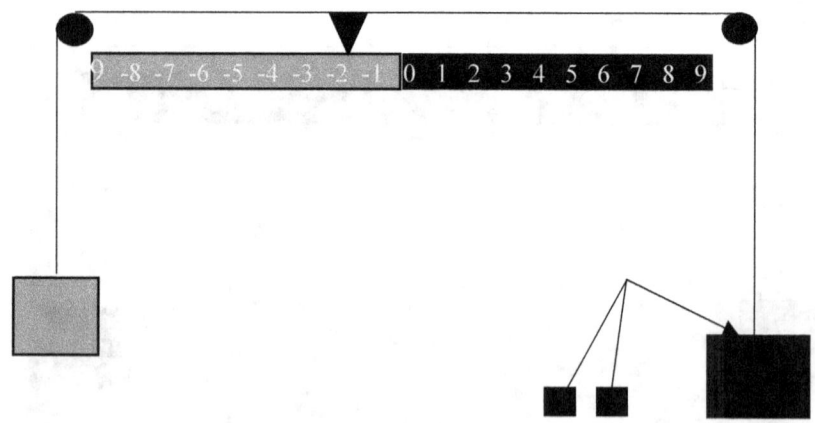

Our pointer has moved to the right by two numbers. It's still on the left-hand side or grey side and it's pointing at the number -2. We've just used our machine to work the problem -4 + (+2) = -2. We'll take the 2 black cubes out of the black bucket the pointer goes back to -4.

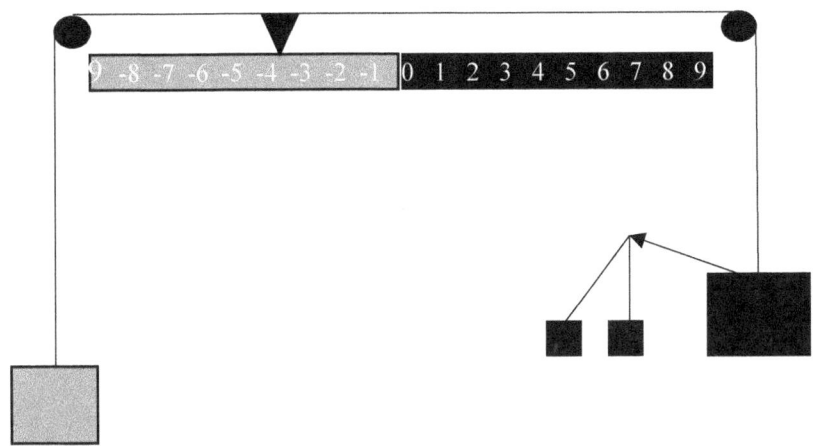

Now we'll add 2 grey cubes to the grey bucket. The pointer moves left to -6. The problem we just worked is -4 + (-2) = -6.

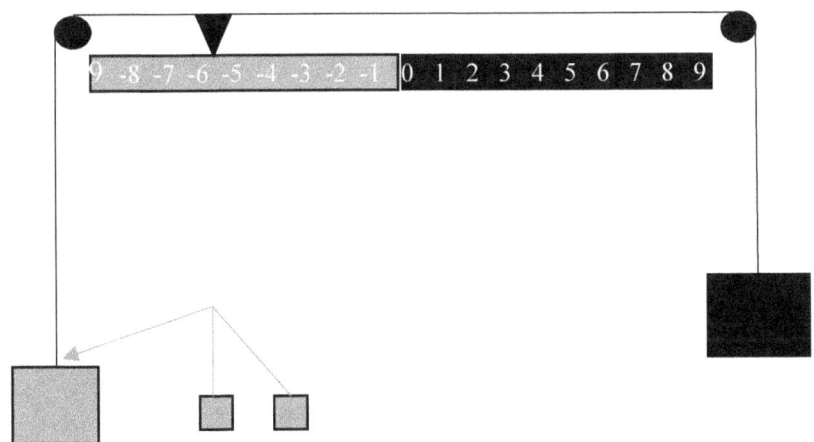

If we write these problems out our usual way they look like this:

$$
\begin{array}{ccccc}
2 & 5 & 5 & -4 & -4 \\
+(+3) & -(+1) & +(-1) & +(+2) & +(-2) \\
\hline
5 & 4 & 4 & -2 & -6
\end{array}
$$

The top row of numbers tells us where the pointer is when we start the problem. **If there is no sign beside the number then it is assumed to be positive** and on the right-hand side of the crossarm. If there is a minus sign beside the number then it is a negative number on the left-hand side or negative side of the crossarm. In the second row of numbers the first sign we come to as we go from left to right tells us if we are adding cubes to a bucket, +, or taking them out, -. The sign and number inside the parenthesis tells us which bucket we're working with, + for black (positive) and – for grey (negative), and how many cubes we are putting in or taking out. The bottom row of numbers tells us where the pointer ends up which is our answer. As was mentioned earlier + and – are both operators and signs for numbers and that frequently mathematicians leave off the + if it's being used as a number sign. For now, we'll continue to show both signs until we get more familiar with how they are used.

Here are 4 more problems we'll work on the number machine:

$$
\begin{array}{cccc}
6 & 6 & 6 & 6 \\
+(+3) & +(-3) & -(+3) & -(-3)
\end{array}
$$

OK, we'll start by putting 6 black cubes in the black bucket. The pointer goes to the +6 on the right-hand side of the crossarm.

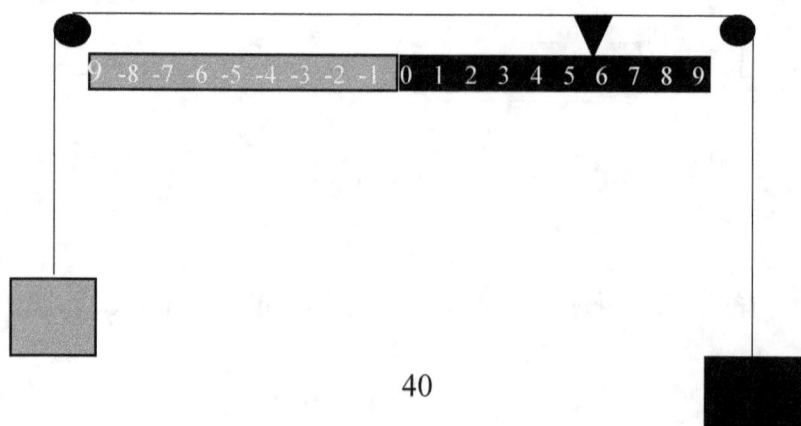

40

For the first problem the first + means we have to put cubes into one of the buckets. The +3 inside the parentheses tells us it will be 3 black cubes in the black bucket.

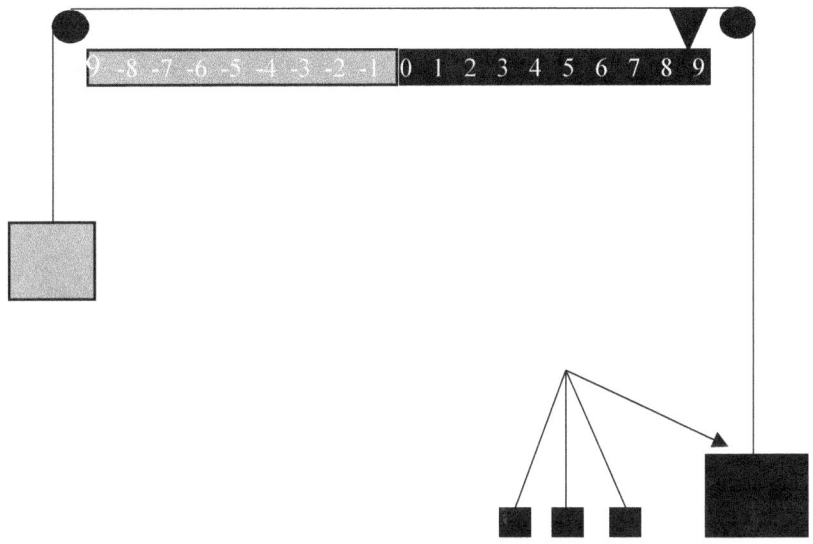

When we do that the pointer moves right to the number 9 so the answer to our first problem is 9.

Let's take 3 cubes out of the black bucket so the pointer is again pointing at the number 6.

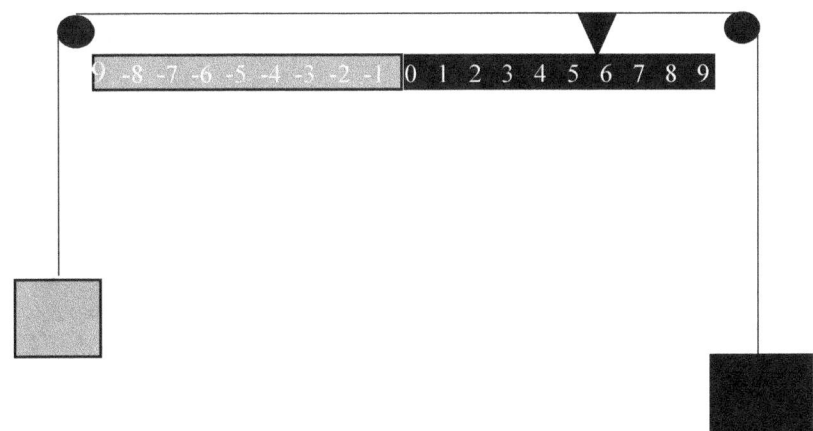

In the second problem, the first sign is + so we'll be adding cubes to a bucket again. This time the minus sign, -, tells us we'll be adding to the grey bucket on the left and we'll be putting in 3 grey cubes.

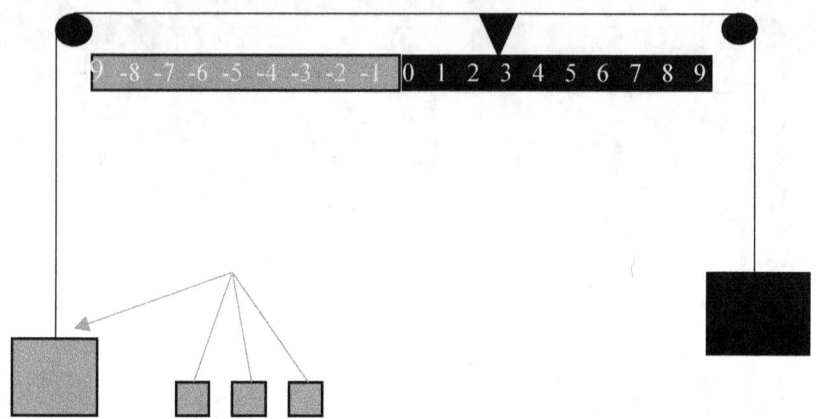

This time the pointer moves left to the number 3 on the right-hand side of the crossarm so our answer is 3.

Let's take the 3 cubes out of the grey bucket so the pointer goes back to the number 6.

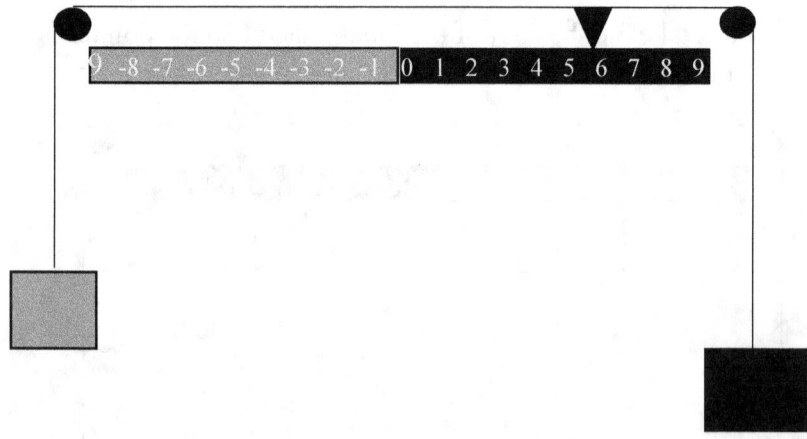

In the third problem the minus sign is telling us we need to take cubes out of a bucket. The + sign and the 3 tells us we need to take out 3 black cubes from the black bucket. The pointer again moves to the number +3.

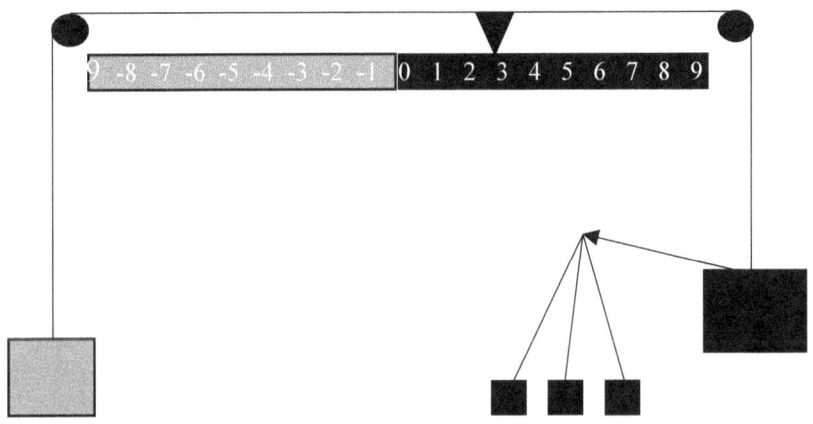

Let's put the cubes back in the black bucket so the pointer once again points at the number 6.

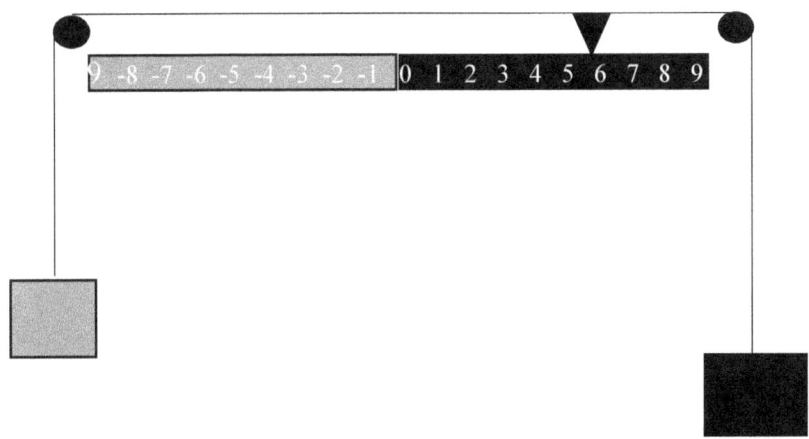

The fourth problem is a little tricky. The first minus sign is telling us we need to take some cubes out of a bucket. The next minus sign, though, is telling us it must be from the grey bucket, but there are no cubes in the grey bucket. What can we do? Well, our number machine not only looks very much like a balance scale, it acts like one. In other words, if we put 3 grey cubes in the grey bucket **and** 3 black cubes in the black bucket the pointer won't move and that will give us some cubes in the grey bucket we can then take out.

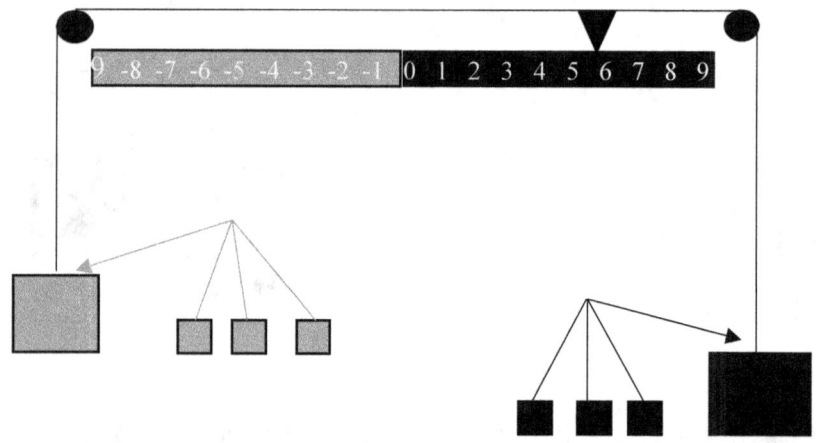

OK, let's take the 3 grey cubes out of the grey bucket and the pointer moves right to the number 9.

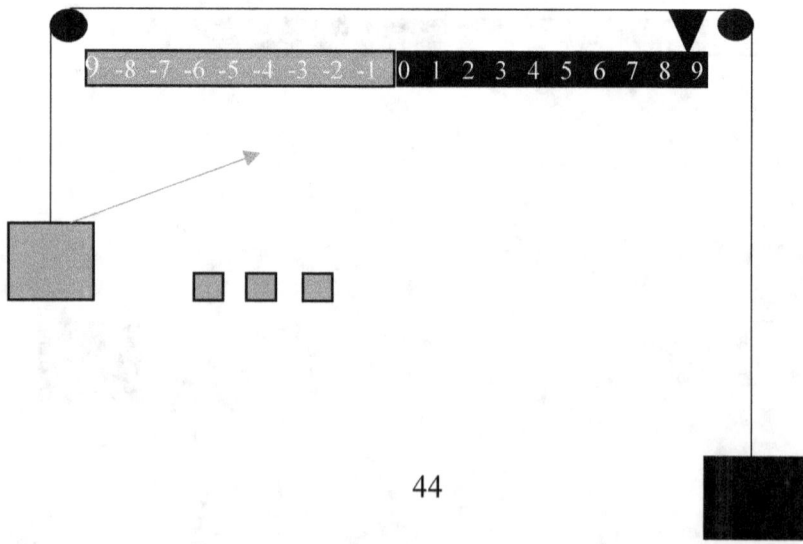

So, what did we get for answers?

6	6	6	6
+ (+3)	+ (- 3)	- (+ 3)	- (- 3)
9	3	3	9

I think I'm starting to see a pattern here. Notice in the first problem both signs in the second row are + and in the last problem both signs are – and the answers are the same as if we had just added the two numbers. In the second and third problems both signs are different and their answers are the same as if we had just subtracted the two numbers.

Conclusion: if the signs are the same, add, if the signs are different, subtract.

This is one of the most important rules to learn at this point in your math career. No matter how far you go in math the idea of like signs being positive and unlike signs being negative will continue to crop up. It is not only fundamental to how our system of mathematics works it is also fundamental to how we use mathematics to solve problems.

A word about our method of solving negative number problems. We showed that we can put the larger number on top, subtract and bring down the sign of the number on top. While this method works, it is not the only method for solving negative number problems. Actually, none of the methods for solving any of the problems in this book are the only methods for solving problems. The attempt here is to show methods that are as simple, versatile and intuitive as possible and that help reinforce the mathematical concepts being taught.

Now that we're a little more familiar with using negative numbers we need to ask ourselves where these things might be used in the real world. We've already seen that we can use them when dealing with money and marbles but it turns out that negative numbers can also be very useful when thinking about altitude. If a plane is going up we can say it has a positive rate of climb, if it's going down then it has a negative rate of climb. If something is going away from us we might say it has a positive velocity meaning it's going away but if it's coming toward us it has a negative velocity. These concepts are frequently used in the military. In physics, the use of negative numbers to indicate direction is very useful. In chemistry, some particles are positively charged while others are negatively charged so negative numbers are used to help figure out how chemicals will react with each other. Getting comfortable with using negative numbers now will pay big dividends in the future.

Exercises:

-16	7	11	-3	-13
-(-9)	-(+5)	+(+3)	-(-9)	+(-7)

21	-31	-18	6	41
+(+9)	-(-9)	+(-1)	-(+8)	+(-11)

-108	74	-10	-10	5
-(+59)	-(+55)	+(-21)	+(+21)	-(-5)

Chapter 5

Multiplication

Multiplication is frequently called a kind of fast addition. For purposes of our book this statement is very true, however, multiplication will not work with all addition problems. Do you remember problems 4 and 5 at the end of chapter 2? Problem 4 was about how many chairs there were at the 5 tables in a restaurant. Problem 5 was about how many wheels were on 12 cars. In both cases we were adding the same number several times. For the chairs, we had 4 chairs at each table and there were 5 tables so we added five 4's together to get 20 chairs. With the cars, we had 4 wheels on each car and there were 12 cars so we added twelve 4's together to get 48 wheels. Addition problems like these can be handled much faster and easier by using multiplication.

To get good at multiplication some memorization will be necessary. This is another of those times when flash cards can come in handy, however, simply working problems can also get you comfortable with this process. So, what is this process of multiplication?

Let's say we want to count how many wheels there are on 2 cars. We count 4 wheels on each car and we have 2 cars so we can use addition and calculate the number of wheels:

$$\begin{array}{r} 4 \\ +4 \\ \hline 8 \end{array}$$

Let's look at this problem again and see if we notice something. Remember way back at the beginning of the book it was mentioned that the human mind likes to group things together? Well, here's a good case of that. Notice we have two groups of 4 wheels. Since we've been practicing our addition we can easily add 4 plus 4 and get 8, but is there another way to do this? Could we,

somehow, write down 2 groups of 4 and then the answer, 8?
Multiplication is that way. Here's what it looks like:

$$
\begin{array}{r}
4 \\
\underline{\times 2} \\
8
\end{array}
$$

The 4 tells us how big the group is. The 2 tells us how many
groups we are adding together. The "x" is called the multiplication
operator sign just like the "+" was the addition operator sign and the
"-" was the subtraction operator sign. Another way to look at this
problem is to say we have a group of 4, now, how many times are
we adding that group to itself? The answer is 2 times so we would
say, 4 times 2 equals 8.

Let's try another problem. How many wheels are there on 3
bicycles? We'll assume that each bike only has two wheels. If we
use addition we would write:

$$
\begin{array}{r}
2 \\
+2 \\
\underline{+2} \\
6
\end{array}
$$

If we use multiplication we would start by saying we have 3
groups of 2 or 2 times 3. It looks like this:

$$
\begin{array}{r}
2 \\
\underline{\times 3} \\
6
\end{array}
$$

This example shows us that multiplication can take less
writing and can be a little faster. What do we do, though, when our
answer is greater than 10, for example that tables and chairs problem
mentioned at the beginning of the chapter? Here we have a group of
4 added together 5 times or 4 times 5. Since we solved this problem
before using addition we know that the answer is 20, but how do we
write it down in our bookkeeping system?

When we used addition to solve this problem we added up our 5 groups of 4 and got 20 or 2 groups of 10 with no 1's left over. We wrote 0 in the 1's column and carried 2 over to the top of the 10's column. We then added the 2 to all of the numbers in the 10's column, but since there were no other groups of 10 we brought the 2 down and there was our answer.

It turns out that multiplication works the same way. Here's what it looks like:

$$\begin{array}{r} 2 \\ 4 \\ \underline{\times\ 5} \\ 20 \end{array}$$

Here are a few more problems to help you get used to using our bookkeeping system with multiplication:

7	8	4	9	5
x3	x2	x6	x3	x7

3	2	6	3	7
x7	x8	x4	x9	x5

Notice that the bottom row of problems are the same numbers as the top row of problems only turned upside down. This was done intentionally. Remember when we learned addition we worked some problems to show that it didn't matter which order we added the numbers. This was called the commutative law. It turns out this same law applies to multiplication. We can multiply 7 times 3 or 3 times 7 and still get 21.

You can easily prove this for yourself using toothpicks. Get out 21 toothpicks. Create groups of 7. How many groups did you

49

get, 3? Good, now put all the toothpicks together again and create groups of 3. How many groups did you get this time, 7? Good!

So far, our problems have involved just numbers in the 1's column, now let's look at something a little bigger. Suppose we wanted to know how many there were in 3 groups of 23. In our bookkeeping system, we would write this:

$$
\begin{array}{r}
23 \\
\underline{\times 3}
\end{array}
$$

We can handle this problem very much the same way we did addition. We'll start with the right-hand column, which in this case is the 1's column and multiply 3 times 3, which is 9. We write the 9 down in the 1's column below the line then move on to the 10's column and multiply 2 times 3, which is 6. We write the 6 below the line in the 10's column and we discover that 23 times 3 equals 69.

Remember when I said mathematicians like to look for patterns? Here's a great example of that. In addition, as the numbers got bigger we had more columns to the left. To add these bigger numbers, we started by adding the numbers in the column on the right then moved to the next column to the left and then the next column to the left and so on until we were finished. Multiplication follows the same pattern. Let's try some bigger numbers and see how it works.

$$
\begin{array}{r}
424{,}312{,}234 \\
\underline{\times\ 2} \\
848{,}624{,}468
\end{array}
$$

To work this problem, we start at the right-hand column and multiply 4 times 2 and get 8. We then move left to the 10's column and multiply 3 times 2 and get 6. We follow this same pattern until we run out of columns with numbers and we have our answer. Here's a few more problems to help you practice.

213	4,142	13	2,311	2,221
X3	x2	x3	x3	x4

24,341	121,312	2,231,131	22,111,222
X2	x3	x3	x4

Let's look at another pattern from addition and see if it works in multiplication. When we added a column of numbers and the total was 10 or more we wrote the leftovers in the column and carried the groups over to the next column to the left, correct? It turns out we can do the same thing in multiplication. Remember, multiplication is just a faster way of adding so it makes sense that the same rules apply. Let's try an example.

$$\begin{array}{r} 25 \\ \underline{\times 3} \end{array}$$

For this problem, we start in the right-hand column and multiply 5 times 3, which is the same as 5 plus 5 plus 5, which is 15. We write the leftovers, 5, in that column and carry the groups, 1, to the top of the next column to the left. Now we multiply 2 times 3 and get 6 then **ADD** the groups we carried over, 1, to get 7 and write that in that column below the line and we have our answer, 75. When we write it out it looks like this:

$$\begin{array}{r} 1 \\ 25 \\ \underline{\times 3} \\ 75 \end{array}$$

Let's try a few more problems to make sure we've got the hang of it.

24,341	121,312	2,231,131	22,111,222
x5	x4	x6	x5

6581	738	295,834	56,136
x6	x5	x9	x7

Notice that by using this system the largest numbers you'll ever have to multiply at one time are 9 times 9. This would be a great time to use some flash cards to memorize what are called the multiplication tables. The tables you have to learn are all of the numbers, 1 through 9, times each other, in other words:

1	2	3	4	5	6	7	8	9
x1	x1	x1	x1	x1	x1	x1	x1	x1

1	2	3	4	5	6	7	8	9
x2	x2	x2	x2	x2	x2	x2	x2	x2

1	2	3	4	5	6	7	8	9
x3	x3	x3	x3	x3	x3	x3	x3	x3

1	2	3	4	5	6	7	8	9
x4	x4	x4	x4	x4	x4	x4	x4	x4

1	2	3	4	5	6	7	8	9
x5	x5	x5	x5	x5	x5	x5	x5	x5

1	2	3	4	5	6	7	8	9
x6	x6	x6	x6	x6	x6	x6	x6	x6

1	2	3	4	5	6	7	8	9
x7	x7	x7	x7	x7	x7	x7	x7	x7

1	2	3	4	5	6	7	8	9
x8	x8	x8	x8	x8	x8	x8	x8	x8

1	2	3	4	5	6	7	8	9
x9	x9	x9	x9	x9	x9	x9	x9	x9

At first this seems like a **lot** of memorization but, thanks to the commutative law, it's not as bad as it looks. Think about it, what's the difference between 5 times 4 and 4 times 5? The answer is, nothing. Because of that we can eliminate almost half of the table and this is what we're left to learn:

1	2	3	4	5	6	7	8	9
x1	x1	x1	x1	x1	x1	x1	x1	x1

	2	3	4	5	6	7	8	9
	x2	x2	x2	x2	x2	x2	x2	x2

		3	4	5	6	7	8	9
		x3	x3	x3	x3	x3	x3	x3

			4	5	6	7	8	9
			x4	x4	x4	x4	x4	x4

				5	6	7	8	9
				x5	x5	x5	x5	x5

					6	7	8	9
					x6	x6	x6	x6

```
   7        8        9
  x7       x7       x7
 ____     ____     ____


            8        9
           x8       x8
          ____     ____


                     9
                    x9
                   ____
```

As was mentioned at the beginning of the book, all learning begins with memorization. The trick is that by frequently using what we've memorized and understanding what it means, the things we memorize quickly become part of our knowledge base. Try to go beyond 8 times 8 equals 64 by continually reminding yourself that what you're really saying is 8 groups of 8 things add up to 64 things. To really get good at this, look around you and try to find places to apply your new knowledge. For example, if you have a tiled floor in your house pick out an area and count how many tiles there are in one row then count the number of rows. Multiply the two together to get the total number of tiles in that area, then count the tiles to check your answer. If you try you can find many places to use multiplication all around you. Now let's take our multiplication to the next level.

We're going to learn some new terms. In all of our previous problems the top number was the size of the group. The formal mathematical name for that group is the **MULTIPLICAND**. The number on the bottom was the number of groups we had. That number is called the **MULTIPLIER.** Together they are known as **FACTORS**. Until now all of our multipliers have been single digit numbers. Let's see what happens when the multiplier has 2 or more digits. Here's an example:

```
    36
  x 12
  ____
```

Here our group size, multiplicand, is 36 and we're adding up 12 of these groups, multiplier. Let's look a little more closely at our multiplier. The number 12 is really the number 10 plus the number 2, right? We'll use that fact to work our problem. First, we'll multiply 36 by 2 then we'll multiply 36 by 10 then we'll add our answers together. Here's what it looks like:

$$
\begin{array}{r}
1 \\
36 \\
\times 12 \\
\hline
172 \\
36 \\
\hline
432
\end{array}
$$

When we multiply by 2 we get 6 times 2 which is 12. We write the 2 in the 1's column and carry 1 over to the 10's column, that's the grey 1. Then we multiply 2 times 3 to get 6 and add the 1 we carried to get 7, which we write down in the 10's column. Next, we multiply 36 by the 1 in the 10's column and write the answer below the answer we just got. 1 times 6 is 6, **WHICH WE WRITE DOWN IN THE 10'S COLUMN.** Next, 1 times 3 is 3 which we write down in the 100's column. Lastly, we draw a line below our two products and do a little addition. First, 2 plus nothing is 2, which we write down in the 1's column. Next, 7 plus 6 is 13 so we write the 3 down in the 10's column and carry a 1 over to the 100's column, that's the other grey 1 next to the 7. Lastly, we add the 3 plus the 1 in the 100's column to get 4. Our answer is 432.

It's very important to keep our columns straight when doing problems like this. One way to do that is to use 0's in the right-hand columns we're not using. In the problem we just worked we could put a 0 to the right of the 6 and below the 2. This doesn't change the value of our answer and helps insure our columns don't get mixed up. It would look like this:

$$
\begin{array}{r}
1 \\
36 \\
\underline{\times 12} \\
172 \\
\underline{360} \\
432
\end{array}
$$

Let's do a problem with a little bigger multiplier to make sure we understand how this works. The 0's used as place holders will be grey colored.

$$
\begin{array}{r}
\mathbf{2} \\
2 \\
241 \\
\underline{\times 527} \\
1687 \\
4820 \\
\underline{120500} \\
127007
\end{array}
$$

First, we multiply by 7. 7 x 1 equals 7. Which we write in the 1's column. 7 x 4 is 28. We write the 8 in the 10's column and carry the 2 to the 100's column. That is shown as a grey 2. Then, 7 x 2 is 14 plus the 2 we just carried equals 16. We write the 6 in the 100's column and the 1 in the 1000's column. Next, we multiply by 5. 5 x 1 equals 5 which we write in the 100's column. 5 x 4 equals 20. We write the 0 in the 1000's column and write the 2 at the top of the 100's column. That 2 is shown as bold. Next, 5 x 2 is 10 plus the 2 we carried equals 12. We write the 2 in the 10,000's column and the 1 in the 100,000's column. Lastly, we add our separate multiplication answers together to get the answer, 127,007.

Here are a few more problems to practice on. Remember that using the 0's as place holders is completely your choice. Some people find them useful while others think they get in the way. Try using both methods on some of the problems and see which way works best for you.

254	3182	671	924	319
x255	x313	x448	x122	x667

4892	7335	386	645	229
x366	x5812	x338	x2212	x506

Notice that in the last problem there is already a 0 in the 10's place of the multiplier. This actually makes our job easier. Did you notice a pattern? For every number in the multiplier there is a row of numbers below the line. In the problem where we multiplied 36 times 12 our multiplier, 12, has 2 digits and we had 2 rows of numbers below the line to add together. In the next problem, 241 times 527, our multiplier, 527, has three digits and there are three rows of numbers below the line to add together. This pattern continued with the exercise problems above with the exception of the last problem, 229 times 506.

You could write this problem out following the pattern and you'd have one row of 0's. However, we can save ourselves some time and work by simply writing one 0 directly below the 0 in the multiplier then start multiplying by 5 in the next column of the same row. Here's the way it would look:

$$\begin{array}{r} 229 \\ \times 506 \\ \hline 1374 \\ 1145\ 0 \\ \hline 115874 \end{array}$$

The grey 0 in the second row is actually serves two purposes. First, it shows we multiplied by the 0 in the 10's column of the multiplier and, second, it serves as a placeholder so that when we start multiplying by 5 our answer starts in the correct column.

The last thing we're going to talk about in this chapter is multiplying negative numbers. This is actually, very easy. Remember that multiplication is really a fast way of adding. It makes sense, then, that the patterns, or rules, we saw when learning negative numbers would still work here, and they do. The rules were:

1. If the signs are the same, then the effect is positive.
2. If the signs are different, then the effect is negative.

For purposes of multiplication, and later in division, we'll state the rules as follows:

1. If the signs are the same, then the answer is positive.
2. If the signs are different, then the answer is negative.

Let's try a few examples and see how it works:

6	6	-6	-6
x2	x (-2)	x2	x (-2)
12	-12	-12	12

There is another aspect to multiplying that we haven't talked about. It's how to handle numbers with decimal points. One example would be money. We're going to put that off for right now because in order to really understand what we're doing we first need to learn about division then fractions. So, for now practice as many problems as you can, look around you for ways to use multiplication and keep practicing the multiplication tables until you can do them with ease.

****

Now that you've got some familiarity with multiplying it's time to introduce a few more ways mathematicians use to show multiplication. As you get further along in your math education you'll start seeing problems written horizontally rather than vertically. There are many reasons for doing this, which we won't go into here. The reason for showing you these other notations now is so you'll be familiar with them the next time you see them. Let's look at the different ways we can write the problem 5 times 7.

5 x 7 =
5 ▪ 7 =
(5)(7) =

Each of the above represent a different notation system for saying the same thing, multiply 5 times 7. Later, when we get into division and then fractions we will also introduce different ways of notating each of them. In the meantime, don't let the different notations confuse you, the same concepts apply no matter what notation system is being used.

Chapter 6

Division

One way to look at division is that it is the opposite of multiplication. In multiplication, we took a number of equal sized groups and combined them into one big group. In division, we'll start with one big group and break it down into some number of smaller equal sized groups. Let's look at an example. We'll start with a multiplication problem, 4 times 3.

$$
\begin{array}{r}
4 \\
\times 3 \\
\hline
12
\end{array}
$$

Now, to do division we are going to introduce a new symbol, $\overline{)}$. Let's see how to use it. We're going to divide 12 by 3. This is the way we will write it:

$$
3\overline{)12}^{\,4}
$$

The number inside the symbol, 12, is the group we're starting with and is called the **dividend**. The number to the left of the symbol, 3, is the size of the smaller groups into which we want to divide 12 and is called the **divisor**. The number on top of the symbol, 4, is the number of groups of 3 there are in 12 and is called the **quotient**. Another way to look at this is to say that 3 is the number of groups we want to divide 12 into and 4 is the size of each group. Either way of looking at it is correct. Let's work a few more examples to get more comfortable with the system.

$$
3\overline{)9}^{\,3} \qquad 6\overline{)12}^{\,2} \qquad 9\overline{)18}^{\,2} \qquad 5\overline{)20}^{\,4} \qquad 7\overline{)14}^{\,2}
$$

These problems are simple and straight forward and are just the first step in learning what has been traditionally called long division. We're going to learn this process one step at a time. Some things to look for as we go through this is that long division is, in many ways, the reverse of multiplication and we will be using multiplication and subtraction to work these problems.

When I say that division is the reverse of multiplication I mean that instead of starting at the right side and working left, we will be starting at the left side and working right. Let's look at an example, 213 divided by 3. The problem looks like this:

$$3\overline{)213}$$

The first thing we'll do is ask ourselves how many groups of 3 are there in 2? Obviously, the answer is none, so the next thing we have to ask ourselves is how many groups of 3 are there is 21? The answer is 7. Since 1 is the further number to the right that we've worked with we put the 7 above the 1 like this:

$$\begin{array}{r} 7 \\ 3\overline{)213} \end{array}$$

Next, we multiply 7 times 3 and write our answer directly below the 2 and 1 and subtract our answer from 21, like this:

$$\begin{array}{r} 7 \\ 3\overline{)213} \\ -21 \\ \hline 0 \end{array}$$

Now, we copy the next number to the right down next to the answer to our subtraction problem, like this (shown in grey):

$$\begin{array}{r} 7 \\ 3\overline{)213} \\ -21 \\ \hline 03 \end{array}$$

Now we ask ourselves how many groups of 3 there are in 3? The answer is 1 so we write the 1 above the 3 and to the right of the 7 in the answer:

$$
\begin{array}{r}
71 \\
3\overline{)213} \\
-21 \\
\hline
03
\end{array}
$$

Now, just like before, we multiply our answer, 1, times 3 and write it below the 03 at the bottom and subtract, like this:

$$
\begin{array}{r}
71 \\
3\overline{)213} \\
-21 \\
\hline
03 \\
3 \\
\hline
0
\end{array}
$$

Since there are no more numbers to bring down we've finished our problem and our answer is 71 or, in other words, there are 71 groups of 3 in 213.

Let's work through a couple more problems to make sure we've got the hang of it. This time we'll divide 432 into groups of 9. Here's what the problem looks like:

$$9\overline{)432}$$

Just like in the last problem, the first thing we need to ask ourselves is, how many groups of 9 are there in 4 things and, of course, the answer is none. The next thing we ask is how many groups of 9 are there in 43 things. Recalling our math tables, we know that there are 4 groups of 9 in 36 things and 5 groups of 9 in 45 things. The way we handle this is that we need the largest number of groups of 9 that is not more than 43 so we'll have to go with 4 groups. Next we multiply 4 times 9 and get 36 then subtract 36 from 43 like this:

$$
\begin{array}{r}
4 \\
9\overline{)432} \\
36 \\
\hline
7
\end{array}
$$

Now, continuing to work to the right, we bring down the 2 and put it next to the 7:

$$
\begin{array}{r}
4 \\
9\overline{)432} \\
36 \\
\hline
72
\end{array}
$$

Next, we ask ourselves how many groups of 9 there are in 72 things. Again, from our multiplication tables, we know there are 8 groups of 9 in 72 so we put that answer on the top to the right of the 4 and multiply it times 9 then subtract from 72:

$$
\begin{array}{r}
48 \\
9\overline{)432} \\
36 \\
\hline
72 \\
72 \\
\hline
0
\end{array}
$$

Since we've gone through all of the numbers in the dividend and our remainder is 0, we're done. There are 48 groups of 9 in 432.

Let's do one more together then you can work some practice problems. This time we'll divide 4,636 by 4:

$$
4\overline{)4636}
$$

We begin by asking how many groups of 4 there are in 4 and

the answer is 1:

```
      1
4)4636
   4
  06
```

Now, how many groups of 4 are there is 6, and, again, the answer is 1:

```
     11
4)4636
   4
  06
   4
   2
```

We've written the 1 at the top next to the first 1, multiplied 1 times 4, written that answer below the 6 and subtracted 4 from 6. We now have a remainder of 2 so let's bring down the next number in the dividend, which is 3:

```
     11
4)4636
   4
  06
   4
  23
```

So, how many groups of 4 are there is 23 things? There are 5 groups of 4 in 20 and 6 groups of 4 in 24. 24 is larger than 23 so we'll have to use 5 groups.

```
    115
4)4636
   4
  06
   4
  23
  20
  36
```

5 times 4 is 20 and 23 minus 20 leaves 3. We've brought down the last number in the dividend, 6, and put it next to the remainder 3. Now, how many groups of 4 are there is 36? I think 9 looks good.

$$
\begin{array}{r}
1159 \\
4{\overline{\smash{\big)}\,4636}} \\
\underline{4} \\
06 \\
\underline{4} \\
23 \\
\underline{20} \\
36 \\
\underline{36} \\
0
\end{array}
$$

9 times 4 is 36. 36 minus 36 is 0 and we're done. Here's a few more problems for you to practice on:

$7{\overline{\smash{\big)}\,6356}}$ $8{\overline{\smash{\big)}\,656}}$ $3{\overline{\smash{\big)}\,7215}}$ $5{\overline{\smash{\big)}\,4725}}$

$6{\overline{\smash{\big)}\,7806}}$ $4{\overline{\smash{\big)}\,6292}}$

****

All of the problems we've worked so far have come out evenly or, in other words, there has been a whole number of divisors in the dividend. Let's see how to handle a problem where things don't work out so nicely. What about 362 divided by 6?

$6{\overline{\smash{\big)}\,362}}$

We have 6 groups of 6 in 36 so let's write down that much:

$$
\begin{array}{r}
6 \\
6{\overline{\smash{\big)}\,362}} \\
\underline{36} \\
02
\end{array}
$$

Now we have 2 left over and no more numbers to bring down. What do we do? Well, there are a couple of ways to handle this. We'll look at the simplest first. We ask ourselves how many groups of 6 there are in 2 things and, of course, the answer is 0 so let's write a 0 at the top next to the 6:

```
      60
  6)362
    36
    ‾‾
    02
     0
     ‾
     2
```

Then, 0 times 6 is 0, which we write down below and subtract from 2 which leaves 2. This left over 2 is called the remainder and is written at the top like this:

```
     60r2
  6)362
    36
    ‾‾
    02
     0
     ‾
     2
```

The lower case "r" stands for remainder. This is telling us that there are 60 groups of 6 in 362 with 2 left over. Here are a few more problems to practice on. In this case the answers are included, all you need to do is the rest of the math the same way we've worked the last problem.

```
   156r2        116r1        913r2        70r3
4)626       2)233       5)4567      6)423
```

Notice that the remainder is **ALWAYS LESS** than the divisor. When you think about it that makes sense. If the remainder was equal to or larger than the divisor then you could get another group out of it. This applies not only to the remainder in the quotient, or answer, but also when we multiply a number in the quotient times the divisor and subtract down below. The answer to that subtraction problem is also a remainder and if it is equal to or greater than our divisor then our number of groups in the quotient needs to be larger. Here's an example. Let's say we're dividing 36 by 5. We ask ourselves how many groups of 5 there are in 36. Suppose we answer 6? Here's what we'd get:

$$
\begin{array}{r}
6 \\
5\overline{)36} \\
30 \\
\hline
6
\end{array}
$$

Since our remainder, 6, is greater than our divisor, 5, we need to make our quotient larger. If we change our quotient from 6 to 7 then 7 times 5 is 35 and when we subtract we get a remainder of 1 and we know our answer is correct.

Now let's look at another way to handle problems that don't come out evenly. Back in chapter 2 we introduced the decimal point. We used it to work with values that were less than 1. We can use it again when doing division. Here's how it works. Let's divide 36 by 5 again:

$$
\begin{array}{r}
7 \\
5\overline{)36} \\
35 \\
\hline
1
\end{array}
$$

If we put a decimal point to the right of the 6 in 36 and add 0's going to the right we won't change the value of 36. In other words, 36.00000 is still equal to 36. We'll put in the decimal point and begin by adding one 0 then bring the 0 down next to our remainder1:

$$
\begin{array}{r}
7. \\
5\overline{)36.0} \\
35 \\
\hline
10
\end{array}
$$

Notice we also put a decimal point in the quotient to the right of the 7. Now we have 10 divided by 5. We know we have 2 groups of 5 in 10 so we write a 2 to the right of the decimal point in the quotient, multiply 2 times 5 and write the answer at the bottom:

$$
\begin{array}{r}
7.2 \\
5\overline{)36.0} \\
35 \\
\hline
10 \\
10 \\
\hline
0
\end{array}
$$

So, what is this answer telling us? It's saying that in a group of 36 things there's a little over 7 groups of 5. We'll look at this a lot closer in the next chapter when we learn about fractions in more detail. For right now here are a few practice problems to help you get more familiar with adding a decimal point in division problems.

$$8\overline{)51} \qquad 4\overline{)67} \qquad 5\overline{)37} \qquad 8\overline{)428} \qquad 5\overline{)693}$$

Something to Think About:

We've had a lot of exercise problems in this book so far and there will be more to come plus there are even more in the back of the book in the appendixes. About now you may start asking yourself, with calculators everywhere why should I work all of these exercise problems? I know when I was first learning math that question crossed my mind many times, and we didn't have calculators, we had mechanical adding machines which were harder to come by.

I eventually learned why doing all of the exercise problems was the best thing to do but that knowledge came with some pain. To help you avoid that pain I'll give you the answer. The answer is in the name, **exercise problems**. The reason for working the problems is to exercise your brain. Both your body **and your mind** get stronger when you exercise them. Sports and working out are known as good ways to strengthen your body. Math is one of the best ways known to strengthen your mind. Working these problems helps you strengthen your mind for solving problems of **ALL** kinds in the real world. Math helps make you a better thinker, and that's **NEVER** a bad thing. Not only should you work all of these exercise problems you should also be looking around you for any chance to use your new math skills on things you see every day. As for calculators, they do have a good purpose at this stage of your math education. Work the problems by hand then use a calculator to check your answers. If one of your answers is wrong, rework the problem by hand until you find your mistake. By doing that you will give your mind its best exercise and it will pay you back.

   REMEMBER: Working a problem wrong then reworking it until you find your mistake is actually **more** beneficial than working the problem right the first time.

****

   Now that you're getting familiar with how long division works we're going to introduce a couple of other ways that mathematicians use to show division. We'll be dividing 12 by 6.

$12 \div 6 =$

$12 / 6 =$

   Both of these are horizontal notation systems. The bottom notation can also be written vertically like this:

$$\frac{12}{6} =$$

   In the horizontal notations, the number on the left is the dividend and the number on the right is the divisor. In the vertical system, the top number is the dividend and the bottom number is the divisor. The vertical notation system is worth remembering because we will be using it extensively in the next chapter on fractions.

Exercises:

$16 \div 4 =$ $8 \div 4 =$ $21 \div 3 =$ $21 \div 7 =$

$$\frac{48}{8} = \qquad \frac{48}{6} = \qquad \frac{36}{9} = \qquad \frac{27}{9} = \qquad \frac{18}{2}$$

$$\frac{32}{4} = \qquad \frac{32}{16} = \qquad \frac{38}{2} = \qquad \frac{54}{9} = \qquad \frac{45}{5}$$

$$\frac{72}{8} = \qquad \frac{63}{7} = \qquad \frac{60}{12} = \qquad \frac{60}{3} = \qquad \frac{60}{4}$$

In these next problems, show your answer as a simple remainder:

$$3\overline{)13} \qquad 4\overline{)23} \qquad 6\overline{)20} \qquad 5\overline{)17} \qquad 7\overline{)27}$$

In the next problems, show your answer as a decimal:

$$4\overline{)14} \qquad 8\overline{)20} \qquad 6\overline{)15} \qquad 12\overline{)27} \qquad 10\overline{)42}$$

Chapter 7

Fractions

There are a number of different ways fractions can be written or expressed and we will look at the two most common but let's start with a definition. What is a fraction? The most basic definition is, **a fraction is part of a whole**. Here's an example.

Suppose there is one piece of cake left and you and a friend decide to share it. You cut it into 2 equal parts and you each take 1. The piece of cake represents the whole thing you're starting with and each of you now has a part or portion of that whole or in other words you each have half of the piece you started with. Here is one way we can write that:

$$\frac{1}{2}$$

This form of notation is sometimes called a **common fraction**. The number 2 on the bottom is called the **denominator** and tells us how many equal parts or portions the piece of cake was cut into. The 1 on top is called the **numerator** and tells how many of those equal portions each of you got.

Here's another example. Suppose you had 2 friends with you and you have a candy bar. You divide it into 3 equal parts and each of you gets 1 part. What fraction of the candy bar would each of you have?

Using the common fraction way of writing fractions, we start with the top number, the numerator, which tells us how many equal parts each of you got, which is 1. The bottom number, the denominator, tells us how many equal parts you divided the candy bar into. The fraction looks like this:

$$\frac{1}{3}$$

Each of you got one third of the candy bar. Just for fun, let's suppose one of your friends decides he shouldn't be eating candy right now and gives you his piece. Now, how much of the candy bar do you have? Good sense tells us that if we had one third to start with and someone gives us another one third then we must now have two thirds and that is correct. Here's what it looks like written in common fraction form:

$$\frac{1}{3} + \frac{1}{3} = \frac{2}{3}$$ ■ + ■ = ■

Notice that we only added the numerators. The denominators are still 3 because we still only have 3 parts of the candy bar. If you now decided to give your other friend one of your pieces how much of the candy bar would you have? If you started out with two thirds of the candy bar and gave away one third good sense would again tell you that you now have one third left and, again, you would be correct. Here's what that looks like:

$$\frac{2}{3} - \frac{1}{3} = \frac{1}{3}$$ ■■ - ■ = ■

Again, all we subtracted were the numerators, the denominators stay the same. This is a **VERY IMPORTANT RULE** when it comes to adding and subtracting fractions:

RULE: **When adding or subtracting fractions all of the fractions must have the same denominator.**

This is called having a **COMMON DENOMINATOR**.

Now, suppose we have a dozen, 12, cookies and we want to divide them evenly between you and your two friends. Since we know how to divide we can easily figure this out, we just divide 12 by 3 like this:

$$\begin{array}{r} 4 \\ 3\overline{)12} \\ \underline{12} \\ 0 \end{array}$$

Therefore, each of you will get 4 cookies, but let's think about this for a minute. If we wrote this as a common fraction we could say that the **whole** dozen was divided into 12 even parts and each of you got 4 parts so our fraction would look like this:

$$\frac{4}{12}$$

but, we also know that since the dozen was divided into three equal portions you each got one third of the dozen or:

$$\frac{1}{3}$$

Since both lines of thinking are true we're left with this situation:

$$\frac{4}{12} = \frac{1}{3}$$

How do we handle it? We're going to learn how to do something called **REDUCING** a fraction. Reducing a fraction is an operation where we divide both the numerator AND the denominator by the same number again and again until there is no longer one number that will divide into both. This can usually be done in one or two steps. Let's get started.

We start by asking ourselves if there is a number that we know will divide into both the numerator and the denominator evenly. Since we just did the long hand division above we know that 4 goes into 12 evenly and, of course, 4 will go into 4 one time. If we divide 4 into the numerator we get 1. If we divide 4 into 12 we get 3, therefore, four twelfths reduces to one third. This is also sometimes called reducing a fraction to its lowest common denominator.

Suppose we didn't recognize right away that 4 would divide into both 4 and 12 what could we have done? Here are some tricks-of-the-trade that can come in handy at times like these.

1. If both the numerator and denominator are even numbers then you can divide both by 2
2. If both the numerator and denominator end in either 0 or 5 then you can divide both by 5
3. If the sum of the digits in the numerator is divisible 3 AND the sum of the digits in the denominator is divisible by 3 then the fraction can be reduced by dividing both by 3.

If we look our fraction of 4/12 we see that both numbers are even so let's divide both by 2. We get the fraction 2/6. When we look at our new fraction we can see that both 2 and 6 are divisible by 2 so let's divide by 2 again. This time we get 1/3. Since neither 1 or 3 are divisible by 2 we are finished.

Now, let's look at the second trick-of-the-trade. We've worked a lot of addition and multiplication problems by now and, if we were looking for patterns, we would have noticed that if we start with 5 and add 5 we get 10, which ends in 0. If we add 5 to 10 we get 15, which ends in 5. If we keep this up we'll eventually notice that as you keep adding 5 the last number in the answer alternates back and forth between 0 and 5. Because this pattern keeps going on forever we can say with certainty that any number that ends in either 0 or 5 can be evenly divided by 5. Look at the following fraction:

$$\frac{25}{30}$$

The numerator ends in 5 and the denominator ends in 0 so both of them can be divided by 5. When we divide 25 by 5 we get 5 and when we divide 30 by 5 we get 6. Our fraction reduces to

$$\frac{5}{6}$$

Our new numerator can still be divided by 5 but since our new denominator can't, we are through.

So, what about our third trick-of-the-trade, how does that work? We'll start with the fraction

$$\frac{27}{57}$$

We can start with either the numerator or denominator, it doesn't matter. Let's start with the numerator. We add the numbers in the numerator together, 2+7. Our answer is 9. We then ask ourselves, is 9 divisible by 3 and, of course it is. Next, we add the numbers in the denominator together, 5+7. The answer is 12 which is also divisible by 3, but let's say we didn't know 12 was divisible by 3. Our answer of 12 still has two digits in it so we can add them together and 1+2=3, which is divisible by 3. Now that we know both the numerator and denominator are divisible by 3 let's reduce our fraction and see what we get.

We can divide 27 by 3 and get 9. We can divide 57 by 3 and get 19. 9 can be divided by 3 again but 19 can't so our fraction is reduced as far as it can be and our answer is

$$\frac{9}{19}$$

Let's work a few problems to make sure we've got a handle on fractions.

1. You and your friend have 63 marbles all together. After playing marbles most of the morning you stop to count your marbles to see how many you have. You count 39 marbles. What fraction of the total marbles do you have? You do not have to reduce your answer.
2. As you're walking down the street one day you start counting

parked cars. As you pass cars you notice that quite a few of them are red colored so you also start keeping track of the number of red cars. When you finish your walk, you have counted 24 parked cars and 15 of them are red. What fraction of the parked cars are red? Again, you do not have to reduce your answer.

3. Reduce your answer to problem 1 to its lowest common denominator.

4. Reduce your answer to problem 2 to its lowest common denominator.

5. A box of cookies says there are 60 cookies in the box. You eat 8 of the cookies and your friend eats 12. What fraction of the box of cookies did you and your friend eat all together? Reduce your answer.

6. You and two friends decide to see who can read the most books in three months. At the end of three months you get together to see how each of you has done. You've read 9 books. One of your friends has read 11 books, and the other friend has read 6 books. What fraction of the total books has each of you read? Reduce your answers if necessary.

7. Reduce the following fractions to the lowest common denominator:

$$\frac{9}{36} \qquad \frac{12}{60} \qquad \frac{4}{24} \qquad \frac{10}{75} \qquad \frac{18}{54}$$

Next, we're going to look at adding and subtracting fractions in a little more detail. We mentioned before that to add or subtract fractions both fractions had to have the same denominator. But, how do we add two fractions that **don't** have the same denominator? Let's imagine we're in the kitchen and we need to put ½ cup of milk in the mixing bowl now and then add another ¼ cup of milk later. Rather than make two trips to the refrigerator we would like to add ½ to ¼ and get that much milk on the first trip.

There was another reason for learning to reduce fractions to the common denominator. We're going to use the reverse process to get a common denominator for two fractions so we can add them. Here's how it works. When reducing a fraction to its lowest common denominator we divided both the numerator and the denominator by the same number. To get a common denominator for two fractions we will **multiply** both the numerator and denominator of a fraction by the same number. We start by looking at the denominators of both fractions:

$$\frac{1}{2} \qquad \frac{1}{4}$$

Notice that if we multiply the first denominator, 2, by 2 we get 4. Remember, if we multiply the denominator by 2 we must also multiply the numerator by 2. Here's what it looks like:

$$\frac{1 \times 2 = 2}{2 \times 2 = 4}$$

We've just given both fractions the same denominator, 4, now we can add them:

$$\frac{2 + 1 = 3}{4 \quad 4 \quad 4}$$

The last step is to check to see if we can reduce our answer, which we can't. So, if we go to the refrigerator and get ¾ cup of milk we'll have enough milk for what we're doing.

Finding a common denominator in that problem was pretty easy and straight forward, but what do we do if the denominators aren't so nice? Look at these two fractions:

$$\frac{4 + 2 =}{9 \quad 5}$$

If we are only adding two fractions together, one trick that always works is to multiply each fraction by the denominator of the other fraction. In this case we would multiply 5/9 by 5/5 and 3/5 by 9/9. This is what it would look like:

$$\frac{4 \times 5}{9 \times 5} + \frac{2 \times 9}{5 \times 9} =$$

This gives us the resulting addition problem:

$$\frac{20}{45} + \frac{18}{45} = \frac{38}{45}$$

Again, check to see if our answer can be reduced, which it can't, so we're finished.

A word of caution here. The method we just used to find a common denominator does work all of the time but, sometimes, it can result in more work for us. Look at the first problem we did, ½ + ¼. If we used the last method to find a common denominator we would have multiplied ½ by 4/4 and ¼ by 2/2. The resulting problem would have been 4/8 + 2/8. Our answer is 6/8 which we would then have to reduce to ¾, the same answer we got the first time. As you can see solving the problem this way works but there are more calculations for us to make. Finding a common denominator by recognizing that you can multiply just one of the fractions by a number, like we did in the first problem, is the best way to solve the problem. Being able to see those simple numbers comes with experience.

The next thing we need to look at is what do we do when we end up with an answer where the numerator is larger than the denominator? Until now our answers have always had a numerator that was less than the denominator. Let's look at this situation: You have 2 candy bars and each one is divided into 3 equal parts like the illustration below:

Now, if you set 4 of the parts aside what fraction do you have set aside?

$$\blacksquare \; + \; \blacksquare \; + \; \blacksquare \; + \; \blacksquare \; = \; \frac{4}{3}$$

As we can see from the illustration we have 4/3 set aside. In the world of fractions when the numerator is smaller than the denominator it is called a proper fraction. When the numerator is greater than the denominator we call it an improper fraction. From the stand point of mathematics there is nothing wrong with improper fractions, in fact, as you get further along in your math studies you'll find there are certain branches of math where improper fractions are very commonly used. However, that having been said, in many circumstances the human mind understands proper fractions faster and easier than improper fractions.

If we look at our example, 4/3, good sense tells us that 3/3's is the same thing as 1 whole candy bar so 4/3's must be 1 whole candy bar plus an additional 1/3 candy bar or 1 1/3. This is a fairly obvious example so the question now becomes; how do we do this for more complicated fractions?

We've already answered this question back when we were learning long division. Let's setup the fraction 4/3 as a long division problem.

$$3\overline{)4}$$

Now, here is the problem worked out:

$$
\begin{array}{r}
1r1 \\
3\overline{)4} \\
3 \\
\hline
1
\end{array}
\; = \; 1 \;\; 1/3
$$

The 1 on the left of the "r" in the answer is telling us how many whole candy bars we have and our remainder of 1 is telling us how many thirds we have left over. Let's try a couple more problems to make sure we've got a handle on this.

$$\frac{10}{4}$$

$$\frac{2\,r\,2}{4)\overline{10}} \quad = \quad 2\,2/4 \;=\; 2\,\tfrac{1}{2}$$

$$4)\overline{10}$$
$$\underline{8}$$
$$2$$

This shows how we use long division to reduce the improper fraction of 10/4's to 2 ½. Let's try one with larger numbers:

$$\frac{230}{15}$$

$$\frac{15\,r\,5}{15)\overline{230}} \qquad = 15\ 5/15 \;=\; 15\ 1/3$$
$$\underline{225}$$
$$5$$

Don't forget that last step of reducing your fraction to its lowest common denominator. Coming up are some problems in adding fractions. Problems like these can often result in improper fractions for answers.

Exercises:

$$\frac{3}{4}+\frac{1}{3} \qquad \frac{2}{3}+\frac{3}{6} \qquad \frac{1}{4}+\frac{3}{4} \qquad \frac{3}{5}+\frac{3}{10} \qquad \frac{7}{12}+\frac{1}{2}$$

$$\frac{3}{16}+\frac{5}{8} \qquad \frac{5}{8}-\frac{3}{16} \qquad \frac{7}{8}+\frac{1}{4} \qquad \frac{3}{4}-\frac{1}{3} \qquad \frac{1}{8}-\frac{1}{4}$$

Concept Problem:
You're in the kitchen helping your mother make cookies. There is a sack with 1 ½ cups of sugar in it. Your mother says she'll need ¾ cup of sugar for this recipe and wants to know how much sugar will be left in the bag after that.

Chapter 8
Decimal Fractions

The next thing we're going to look at are what's called decimal fractions. To do this we're going to start by looking at another way to handle long division, but first let's review something we learned about back in chapter 2, money.

When we talked about money we introduced something called the decimal point. Remember that as we move from right to left each column is 10 times larger than the previous column. Another way to look at this is that as we move from left to right each column is only 1/10 the size of the previous column. This pattern continues no matter how many columns there are. The purpose of the decimal point is to show us which column is the one's column. In other words, the column just to the left of the decimal point is the one's column and the column just to the right of the decimal point is the 1/10 column. Look at the example below.

100's	10's	1's		1/10th	1/100th	1/1000th
1	1	1	.	1	1	1

If we look at the number 12, for example, we can see that we can add as many 0's to the left of the number as we want without changing the value of the number: 00000012 is still just 12. Likewise, we can add as many 0's to the right of the decimal point as we want without changing the value of the number: 12.000000 We're going to use this concept with long division to make decimal fractions.

Let's divide 3 by 4:

$$4\overline{)3}$$

Obviously, 3 is too small to have a group of 4 in it so let's put a decimal point on the right side of the 3 and add a 0 to the right of that. At the same time, we'll put another decimal point in the quotient right above the one we just added to the dividend.

$$
\begin{array}{r}
. \\
4\overline{)3.0}
\end{array}
$$

Now, we ask ourselves how many groups of 4 there are in 30. The answer is 7 so we write that just to the right of the decimal point in the quotient and multiply 7 times 4 as usual:

$$
\begin{array}{r}
.7 \\
4\overline{)3.0} \\
2\,8 \\
\hline
2
\end{array}
$$

When we subtract, we have 2 left over, so, let's add another 0 to the right end of the dividend and bring that down next to the 2:

$$
\begin{array}{r}
.7 \\
4\overline{)3.00} \\
2\,8 \\
\hline
20
\end{array}
$$

Now, how many groups of 4 are there in 20? The answer is 5 so we write that in the quotient next to the 7, multiply 5 times 4 and subtract:

$$
\begin{array}{r}
.75 \\
4\overline{)3.00} \\
2\,8 \\
\hline
20 \\
20 \\
\hline
0
\end{array}
$$

Since our remainder is 0 we are finished. We have just converted the proper fraction ¾ into the decimal fraction .75.

Let's look at one more problem then try some exercises. We'll divide 83 by 8.

$$8\overline{)83}$$

Here's the first part of the problem:

```
      10
  8)83
      8
     ──
     03
      0
     ──
      3
```

We have a remainder of 3 so we'll add a decimal point in the dividend with a 0 after it and a decimal point in the quotient to the right of the 0. Next, we'll continue to add 0's to the dividend, bring them down and continue dividing like we did before:

```
       10.375
  8)83.000
      8
     ──
     03
      0
     ──
      3 0
      2 4
      ───
        60
        56
       ──
        40
        40
       ──
         0
```

Exercises:

12/5   45/8   2/4   36/16
74/10

83

Chapter 9
Multiplying and Dividing Fractions

In chapter 5 when we learned about multiplication we saw that multiplying 5 times 2 is the same as multiplying 2 times 5. If we multiply 2 times 5 we are asking, "What is the sum of five twos added together." If we multiply 5 times 2 we are asking, "What is the sum of two fives added together." When we multiply fractions, we are asking ourselves the same question. If we multiply ½ times 4 it's the same as adding four ½'s like this:

$$\frac{1}{2} + \frac{1}{2} + \frac{1}{2} + \frac{1}{2} = \frac{4}{2}$$

The original problem could have been written like this:

$$\frac{1}{2} \times 4 \qquad \text{or} \qquad \frac{1}{2} \times \frac{4}{1}$$

We can write a whole number (integer) as the number over 1 without changing its value. What we're saying is that we have 4 things and we're treating them as 1 group.

Now, if we reduce our answer, 4/2, we get 2. We know that multiplying ½ by 4 is the same as multiplying 4 times ½ so we can conclude that 4 times ½ also equals 2. When we multiply 4 times ½ what we're asking ourselves is, "What is ½ of 4".

Let's try another example. Let's multiply ¼ times 8. Again, we'll start by adding ¼ eight times:

$$\frac{1}{4} + \frac{1}{4} + \frac{1}{4} + \frac{1}{4} + \frac{1}{4} + \frac{1}{4} + \frac{1}{4} + \frac{1}{4} = \frac{8}{4}$$

When we reduce our answer, we get 2. Can we recognize a pattern here? When we multiply a whole number, or integer, by a fraction we multiply the numerator times the whole number and keep the denominator. Let's try one more problem, 3/8 times 8. We'll start by adding 3/8 eight times:

$$\frac{3}{8} + \frac{3}{8} + \frac{3}{8} + \frac{3}{8} + \frac{3}{8} + \frac{3}{8} + \frac{3}{8} + \frac{3}{8} = \frac{24}{8}$$

When we reduce our answer, we get 3. We can see that the pattern works.

Now let's see what happens when we multiply one fraction times another. We'll start by multiplying ½ times ½. We'll use our candy bar again. This time we'll start with ½ of our candy bar.

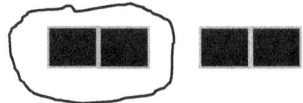

Next, we ask the question, "What is ½ of ½?" The illustration below shows that ½ of ½ of our candy bar is ¼ of our candy bar.

   ½ x ½ = ¼

In this case, it looks like all we did is multiply the denominators together. Let's try a more complex example and see if that pattern holds. We'll multiply ¾ times ½. Here's an illustration of ¾ of the candy bar:

Now we ask ourselves, "What is ½ of these 3 pieces?" The answer is 1 ½ pieces, but what does that mean with respect to the original candy bar? One piece is ¼ of the original candy bar. ½ of one of those pieces is 1/8 of the original candy bar. We can also see that one ¼ piece is the same as two 1/8 pieces so we can add 2/8 and 1/8 to get 3/8. We just showed that ½ of ¾ is 3/8. To get that answer mathematically it looks like this:

$$\frac{3}{4} \times \frac{1}{2} = \frac{3}{8}$$

The patterns from the previous two examples still hold. So, what have we learned? To multiply two fractions we multiply the numerators, which becomes the new numerator, then we multiply the denominators, which becomes the new denominator, then reduce the answer if possible. Another way to say this is, **multiply the tops then multiply the bottoms then simplify**.

If we go back to the first problem we worked, ½ times 4, and look at the way we wrote it the third time:

$$\frac{1}{2} \times \frac{4}{1} = \frac{4}{2} = 2$$

You can see that we are still multiplying the numerators together, then the denominators together, then simplifying. Our rule still works.

OK, let's see how to divide by fractions. We'll start with a simple problem, 1 divided by ½. To understand this problem, we need to ask ourselves the same thing we asked when we first learned how to divide, "What is this problem really asking us?" What we're really being asked here is how many 1/2's are there in 1? I think if you cut a candy bar in half it's pretty easy to see that there are 2 halves in one bar, in other words there are two 1/2's in 1 or 1 divided by ½ is 2.

Here's a little more challenging problem. What is ¾ divided by ½? This question is asking us how many 1/2's are there in ¾? The illustration below shows a candy bar that's been divided into 4 parts with a circle around 3 of them.

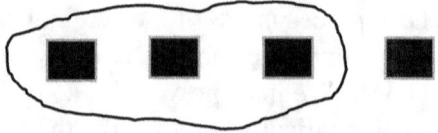

Looking at the 3 parts in the circle we can see that 2 of them equals ½ of the candy bar with one ¼ piece left over. That left-over piece is equal to ½ of ½ of the candy bar. In other words, the picture is telling us that there are 1 ½ halves in ¾ of the candy bar.

The question now is, how do we use our bookkeeping system do work these kinds of problems. The rule we use is, **invert the divisor and multiply**. Here's how the problem looks when we set it up:

$$\frac{\dfrac{3}{4}}{\dfrac{1}{2}}$$

The ¾ on the top is the dividend. The ½ on the bottom is the divisor. To work the problem, we take the ½, turn it upside down or invert it and bring it up on top of the line like this:

$$\frac{3}{4} \times \frac{2}{1}$$

The next step is to multiply the way we just learned and simplify:

$$\frac{3}{4} \times \frac{2}{1} = \frac{6}{4} = 1\,½$$

This also works when dividing whole numbers, or integers, by fractions. Remember integers can be written as the number divided by 1. The first problem we worked, 1 divided by ½, can be worked like this:

$$\frac{\dfrac{1}{1}}{\dfrac{1}{2}} \quad = \quad \frac{1}{1} \times \frac{2}{1} = \frac{2}{1} = 2$$

The last thing to look at is how to handle mixed numbers.

Mixed numbers are numbers where we have an integer and a fraction, like our answer of 1 ½ above. In order to multiply or divide these numbers we must first convert them into an improper fraction. The process of converting them is the reverse of simplifying them. We begin by multiplying the whole number by the denominator, which in this case is 1 x 2. We then add the numerator to that answer or 2 + 1 = 3. This becomes our new numerator and the 2 remains our denominator and our improper fraction is:

$$\frac{3}{2}$$

Let's convert 5 ¾ to an improper fraction.

5 x 4 = 20
20 + 3 = 23

Then use the original denominator of 4 to get $\frac{23}{4}$

Let's do a problem and put all of these new ideas together. We'll divide 4 2/3 by ¾. First, convert 4 2/3 into an improper fraction:

4 x 3 = 12 then 12 + 2 = 14. Now apply the denominator to get $\frac{14.}{3}$

Next, we'll write our problem out as a fraction divided by a fraction:

$$\frac{\frac{14}{3}}{\frac{3}{4}} = \frac{14}{3} \times \frac{4}{3} = \frac{56}{9} = 6\ 2/9$$

Of course, the same method works when dividing one mixed number by another, just convert both numbers into improper fractions and divide a fraction by a fraction.

Exercises:

Multiplication of Fractions:

$$\frac{1}{4} \times \frac{3}{8} \qquad 1\ 3/8 \times 2\ \frac{1}{4} \qquad 1\ \frac{1}{2} \times 3/8 \qquad 2/3 \times 3/5 \qquad 12 \times \frac{3}{4}$$

Division of Fractions:

$$\frac{\dfrac{3}{4}}{\dfrac{1}{2}} \qquad\qquad \frac{2\ \frac{1}{2}}{\dfrac{3}{8}} \qquad\qquad \frac{\dfrac{3}{8}}{\dfrac{3}{4}} \qquad\qquad \frac{1\ \frac{1}{4}}{\dfrac{1}{8}}$$

Concept Question: Your mother is baking cookies. The recipe calls for ¾ cup of flour for each batch. There are 3 1/8 cups of flour left in the bag. How many batches of cookies can she bake with the flour she has left in the bag?

Chapter 10
Conclusion

What is conceptual math? The last question in chapter 9 is an example of conceptual math. Many of the examples throughout the book using toothpicks, marbles, cookies, etc. are all examples of conceptual math. Conceptual math is the way we humans learned math in the first place and how we taught it for over 6,000 years.

Conceptual math is the process by which we look at the world around us and apply numbers to it then use those numbers to answer questions and solve problems.

Right now, the United States is going through a math crisis. The math and science scores of our children are constantly dropping when compared to the rest of the world. Alarmed by this drop many schools began rigging the test results to make students look better. When that failed, they began blaming the students saying they weren't as smart as former students.

So, what really happened? Somewhere, back in the 1960's, someone decided to change how math, science and reading were being taught. It was decided that math would be easier to learn if all the student had to do was memorize a method for solving a given type of problem. Math class was reduced to the students being shown different kinds of problems and the method for solving each kind of problem and told to memorize it all.

This kind of education only exercises the part of the brain that memorizes. While this is not necessarily a bad thing it does eliminate exercising the rational part of the brain. The rational part of the brain is what we use to analyze and solve problems. When it comes to surviving as an individual, a country and a species the rational part of the brain is by far the most important part and, as stated in the introduction, learning conceptual math is the best way to develop and exercise that part of the brain.

The wonderful thing about learning concepts is that it frees you to solve problems the way you want to. There is always more than one way to correctly solve any problem. Everyone's brain is wired a little differently and sees the world through different eyes. The best way to solve a problem is the way that works best with your brain and the way you see the world. The methods I've shown in this book for solving problems are, obviously, not the only methods that work. It can be fun and good mental exercise to see if you can come up with another way of solving these problems. Another good way to invest your time is searching the internet for other articles on math and seeing if you can find other ways of working the problems. Other good sources for math instruction are groups like the Sylvan Learning Centers and the Scottish Rite Masons. Both of these groups teach math using concept based techniques and have great success.

Throughout this book I have used examples of, what we used to call, word problems. Over the years I have also heard them called story problems. Whatever you call them they have one thing in common, most students don't like doing them. The irony is that these are the most important kinds of problems to work. These are the problems that really teach us how to apply the math concepts we have just learned. Additionally, as we get older it becomes apparent that this is the way everyday problems will be presented to us. All around us the world is presenting us with golden opportunities to practice math. The kitchen, the playground, the parking lot, the restaurant, the ball game, the park, walking down the street, all of these offer us many chances to use our math. Welcome these opportunities and take advantage of them whenever you can.

Here's an example of two different ways to solve the same problem. One cold night you and 5 friends get together to watch a movie. You decide to make hot chocolate for everyone. You go to the cupboard get out the mix and discover you have 1 ½ cups left. The directions say it takes ¼ cup of mix for each serving. Do you have enough mix to make hot chocolate for everyone?

One way to solve this problem is to divide the amount of chocolate you have, 1 ½ cups, by the amount of mix you need for each serving, ¼.

91

$$\frac{1\frac{1}{2}}{\frac{1}{4}} \quad = \quad \frac{3}{2} \times \frac{4}{1} \quad = \quad \frac{12}{2} \quad = \quad 6$$

Another way to solve this problem is to multiply the amount of each serving, ¼, by the number of servings you want, 6, and compare that to the amount of mix you have, 1 ½ cups.

$$\frac{1}{4} \times 6 \quad = \quad \frac{6}{4} \quad = 1\frac{1}{2}$$

As you can see, either way shows us what we need to know so either solution is valid. This is when math gets to be fun. Looking for different ways to solve problems exercises our minds in problem solving and also gives us an excuse to look at the world around us from a different perspective than we may be used to.

Following this chapter are some appendixes with exercise problems. Most of these problems are there to help the student get familiar with the notation system shown in the chapters and to get used to working problems using that system. Some word problems have been included to give the student an opportunity to use the concepts talked about throughout the book. Take your time and have fun working the problems.

One last word about math. The concepts taught in this book are fundamental to ALL mathematics. Notation systems change over time, people are always coming up with new and different ways to work problems and, occasionally, someone creates a whole new type of mathematics, but the basics will always apply. In the end you are still counting something. If you get to the point where you study differential calculus or integral calculus you will discover that each step of solving a problem still involves addition, subtraction, multiplication and division. Learn the basics well and it will serve you for your whole life.

Good luck and enjoy!

Appendix A

Addition

8	3	5	9	7	4	1	2	6	1
6	6	2	5	2	7	3	9	1	7
7	1			4		6		8	5
								3	6
									9

15	2	7	234	125,681	1
3	61	2	51	1,193	11
21	13	84	111	17	111
5		22	6	100,000	1,111
					11,111
					87,655

Add the following numbers:
2,515 + 12 + 333 + 46 + 55 =
(hint: rewrite the problem vertically)

Answers to Chapter 2 Questions:

1. Total 38
2. Total 38
3. Sums for question 3: 10, 10, 10, 10, 10, 10, 10, 10, 10
4. 20 chairs
5. 48 wheels

Appendix B

93	727	321	564
-74	-186	-39	-65
19	541	282	499

659	1015	676	641
-361	-9	-577	-550
298	1006	99	91

496	187	288	576
-87	-88	-59	-567
409	99	229	9

5216	665	1115	8251
-317	-6	-6	-252
4899	659	1109	7999

12.03	5.23	17.23	3.12
-4.95	-0.25	-6.04	-2.22
7.08	4.98	11.19	0.9

212.35	89.98	73.12	43.55
-13.46	-1.19	-55.03	-0.36
198.89	88.79	18.09	43.19

9.554	1.892	24.823	12.5
-0.005	-0.793	-1.824	-1.376
9.549	1.099	22.999	11.124

0.587	9.055	4.656	557.3
-0.026	-0.016	-3.998	-0.4
0.561	9.039	0.658	556.9

Appendix C
Negative Numbers

Answers to chapter 4 practice problems:

-16	7	11	-3	-13
-(-9)	-(+5)	+(+3)	-(-9)	+(-7)
-7	+2	+14	+6	-20

21	-31	-18	6	41
+(+9)	-(-9)	+(-1)	-(+8)	+(-11)
+30	-22	-19	-2	+30

Some Practice Problems:

1. $-32 - (-16) = -16$
2. $32 - (-16) = 48$
3. $-16 + 48 = 32$
4. $78 + 21 - (+12) + (-9) = 78$

5.Bill has $5.27 in his pocket. He stops by the convenience store to get a few snacks. He gets a small bag of chips for $2.15, a candy bar for $2.00 and a bottle of water for $2.00. Does Bill have enough money to pay for everything?

$$
\begin{array}{r}
5.27 \\
-2.15 \\
-2.00 \\
\underline{-2.00} \\
-0.88
\end{array}
$$

Since our answer is a negative number that means Bill is $.88 short of having enough to pay for everything.

Appendix D
Multiplication

Answers to Chapter 5 problems:

213	4,142	13	2,311	2,112
x3	x2	x3	x3	x4
639	8,284	39	6,933	8,448

24,341	121,312	2,231,131	22,111,222
x2	x3	x3	x4
48,682	363,936	6,693,393	88,444,888

24,341	121,312	2,231,131	22,111,222
x5	x4	x6	x5
121,705	485,248	13,386,786	110,556,110

6581	738	295,834	56136
x6	x5	x9	x7
39,486	3,690	2,662,506	392,952

254
x255
64,770

3182
x313
995,966

671
x448
300,608

924
x122
112,728

4892
x366
1,790,472

7335
x5812
42,6312,020

386
x388
130,468

645
x2212
1,426,740

Appendix E
Division

Answers to Chapter 6 problems:

```
    908           82          2405          945         1301         1573
7)6356        8)656        3)7215        5)4725       6)7806       4)6292
  63            64             6            45           6            4
  05            16            12            22          18           22
   0            16            12            20          18           20
  56             0            01            25          00           29
  56                           0            25           0           28
   0                          15             0           6           12
                              15                         6           12
                               0                         0            0
```

Appendix F
Fractions

Answers to Chapter 7 problems:

$$\frac{3}{4} + \frac{1}{3} = \frac{9}{12} + \frac{4}{12} = \frac{13}{12} = 1\ 1/12$$

$$\frac{2}{3} + \frac{3}{6} = \frac{4}{6} + \frac{3}{6} = \frac{7}{6} = 1\ 1/6$$

$$\frac{1}{4} + \frac{3}{4} = \frac{4}{4} = 1$$

$$\frac{3}{5} + \frac{3}{10} = \frac{6}{10} + \frac{3}{10} = \frac{9}{10}$$

$$\frac{7}{12} + \frac{1}{2} = \frac{7}{12} + \frac{6}{12} = \frac{13}{12} = 1\ 1/12$$

$$\frac{3}{16} + \frac{5}{8} = \frac{3}{16} + \frac{10}{16} = \frac{13}{16}$$

$$\frac{5}{8} - \frac{3}{16} = \frac{10}{16} - \frac{3}{16} = \frac{7}{16}$$

$$\frac{7}{8} + \frac{1}{4} = \frac{7}{8} + \frac{2}{8} = \frac{9}{8} = 1\ 1/8$$

$$\frac{3}{4} - \frac{1}{3} = \frac{9}{12} - \frac{4}{12} = \frac{5}{12}$$

$$\frac{1}{8} - \frac{1}{4} = \frac{1}{8} - \frac{2}{8} = \frac{-1}{8}$$

Concept problem:

$$1\ \tfrac{1}{2}\ -\ \tfrac{3}{4} = \frac{3}{2} - \frac{3}{4} = \frac{6}{4} - \frac{3}{4} = \frac{3}{4}$$

Appendix G
Decimal Fractions

Answers to Chapter 8 Problems:

$$12/5 = 5 \overline{\smash{)}\ \begin{array}{r} 2.4 \\ 12.0 \end{array}}$$

```
                2.4
12/5  =  5) 12.0
            10
            ──
            20
            20
            ──
             0
```

```
              5.625
45/8  =  8 ) 45.000
             40
             ──
              5 0
              4 8
              ───
               20
               16
               ──
               40
               40
               ──
                0
```

```
             .5
2/4  =  4 ) 2.0
            20
            ──
             0
```

```
             2.25
36/16  =  16) 36.00
              32
              ──
              40
              32
              ──
              80
              80
              ──
               0
```

$$74/10 = 10\overline{)74.0}$$
$$\begin{array}{r} 7.4 \\ 10\,\overline{)74.0} \\ 70 \\ \hline 40 \\ 40 \\ \hline 0 \end{array}$$

Appendix H
Multiplying and Dividing Fractions

Answers to Chapter 9 Problems:

Multiplication of Fractions:

$$1 \times 3 = 3$$

$$1\ 3/8 \times 2\ \tfrac{1}{4} = \frac{11}{8} \times \frac{9}{4} = \frac{99}{32} = 3\ 3/32$$

$$1\ \tfrac{1}{2} \times 3/8 = \frac{3}{2} \times \frac{3}{8} = \frac{9}{16}$$

$$\frac{2}{3} \times \frac{3}{5} = \frac{6}{15} = \frac{2}{5}$$

$$12 \times \tfrac{3}{4} = \frac{12}{1} \times \frac{3}{4} = \frac{36}{4} = 9$$

Division of Fractions:

$$\frac{\tfrac{3}{4}}{\tfrac{1}{2}} = \frac{3}{4} \times \frac{2}{1} = \frac{6}{4} = 1\ 2/4 = 1\ \tfrac{1}{2}$$

$$\frac{2\ \tfrac{1}{2}}{3/8} = \frac{5/2}{3/8} = \frac{5}{2} \times \frac{8}{3} = \frac{40}{6} = \frac{20}{3} = 6\ 2/3$$

$$\frac{\tfrac{3}{8}}{\tfrac{3}{4}} = \frac{3}{8} \times \frac{4}{3} = \frac{12}{24} = \tfrac{1}{2}$$

$$\frac{1\frac{1}{4}}{\frac{1}{8}} = \frac{\frac{5}{4}}{\frac{1}{8}} = \frac{5}{4} \times \frac{8}{1} = \frac{40}{4} = 10$$

Concept Problem:

To answer this question, we must divide the flour that is left by the amount of flour needed for each batch:

$$\frac{3\frac{1}{8}}{\frac{3}{4}} = \frac{\frac{25}{8}}{\frac{3}{4}} = \frac{25}{8} \times \frac{4}{3} = \frac{100}{24} = 4\ 4/24 = 4\ 1/6$$

This answer is telling us that we have enough flour for 4 batches of cookies with some flour left over.

www.ingramcontent.com/pod-product-compliance
Lightning Source LLC
Chambersburg PA
CBHW071214220526
45468CB00002B/601